ATMOSPHERIC TRACER
TECHNOLOGY AND APPLICATIONS

ATMOSPHERIC TRACER
TECHNOLOGY
AND APPLICATIONS

Edited by

Jody H. Heiken

Los Alamos National Laboratory
Los Alamos, New Mexico

Organized and Compiled by

Sumner Barr, William E. Clements, Paul R. Guthals

Los Alamos National Laboratory
Los Alamos, New Mexico

Workshop Chairman

David S. Ballantine

U.S. Department of Energy
Washington, DC

np | **NOYES PUBLICATIONS**
Park Ridge, New Jersey, U.S.A.

Published in the United States of America by
Noyes Publications
Mill Road, Park Ridge, New Jersey 07656

10 9 8 7 6 5 4 3 2 1

Library of Congress Cataloging-in-Publication Data

Atmospheric tracer technology and applications.

     Papers presented at Atmospheric Tracers Workshop held
in Santa Fe, N.M., May 21–24, 1984 and sponsored by the
Office of Health and Environmental Research of the
U.S. Dept. of Energy.
     Includes index.
     1. Environmental chemistry--Congresses.
2. Tracers (Chemistry)--Congresses. 3. Atmospheric
diffusion--Congresses. 4. Air--Pollution--Congresses.
5. Aerosols--Congresses. I. Heiken, Jody H. II. Barr,
Sumner. III. Clements, William E. IV. Guthals,
Paul R. V. Atmospheric Tracers Workshop (1984 :
Santa Fe, N.M.) VI. United States. Dept. of Energy.
Office of Health and Environmental Research.
TD193.A76   1986   628.5'3   86-5266
ISBN 0-8155-1082-9

Foreword

This book presents information on atmospheric tracer technology and applications, based on a workshop held in Santa Fe, New Mexico in May, 1984. The objectives of the workshop were to summarize the status of atmospheric tracer technology, provide information exchange for members of the atmospheric community, define areas for future applications of tracers, and identify requirements for new tracer techniques.

Topics specifically discussed in the book include dry deposition and resuspension, global scale tracers and their applications, acid precipitation, natural sources of inorganic and organic gases and aerosols, and regional air quality.

A major objective of research in atmospheric science over the past several decades has been to obtain an enhanced description of transport, transformation, and deposition of trace constituents present in the atmosphere. Interest in these processes has been heightened with the recognition that pollutant materials transported over long distances may exert significant effects at receptor locations, and the consequent desire to improve knowledge of source-receptor relationships.

Much of the motivation for development and application of techniques employing tracer compounds in the atmosphere has arisen from the desire to understand transport and dispersion processes, and this sort of application is the subject of several papers presented at this workshop. Tracers are employed in such applications for the purpose of identifying particular air parcels and as a means of determining the amount of dilution that has taken place as a function of position and/or time. Tracer compounds useful for these purposes include both deliberately introduced materials and materials that are characteristic of a particular source or source region.

There is a wide range of practical applications for tracer experiments. In many cases the tracer allows the documentation of a potential airborne hazard without actually emitting hazardous material or the evaluation of the expected outcome of an expensive process alteration. Some examples of the former case are testing of some rocket fuels, developing chemical defense strategies, establishing safety procedures for possible accidents in the handling and transportation of hazardous materials, and assessing the fire and explosion hazards present in handling special materials such as liquified natural gas. The practical use of tracers for economic benefit includes designing emissions systems, for both air quality control and dissemination systems such as in forest and crop spraying or cloud seeding, and siting of industrial facilities. Other applications include source attribution, which is currently of interest in the acid rain problem. This book is considered timely, as there have been a number of significant advances over the past 5 years.

The information in the book is from *Atmospheric Tracer Workshop,* held at Santa Fe, New Mexico, May 21–25, 1984; edited by Jody H. Heiken, Los Alamos National Laboratory, INC Division, for the U.S. Department of Energy, December 1984.

The table of contents is organized in such a way as to serve as a subject index and provides easy access to the information contained in the book.

Advanced composition and production methods developed by Noyes Publications are employed to bring this durably bound book to you in a minimum of time. Special techniques are used to close the gap between "manuscript" and "completed book." In order to keep the price of the book to a reasonable level, it has been partially reproduced by photo-offset directly from the original report and the cost saving passed on to the reader. Due to this method of publishing, certain portions of the book may be less legible than desired.

NOTICE

Contents and Subject Index

PART IV
INVITED PRESENTATIONS

PART V
AD HOC PRESENTATIONS

Part I

Introduction

ATMOSPHERIC TRACER WORKSHOP

Organized and Compiled by

Sumner Barr, William E. Clements, Paul R. Guthals

ABSTRACT

   The Department of Energy (DOE) Office of Health and
Environmental Research sponsored an Atmospheric Tracers
Workshop in Santa Fe, New Mexico, May 21-24, 1984. The
objectives of the workshop were to (1) summarize the status of
atmospheric tracer technology; (2) provide an information
exchange forum for members of the atmospheric community;
(3) define areas for future application of tracers; and (4) identify
requirements for new tracer techniques.
   Conference attendees included representatives from the
DOE National Laboratories, National Oceanic Atmospheric
Agency (NOAA), industry, academic groups, and government-
sponsored organizations who are interested in atmospheric
dynamics. The workshop opened with a series of invited overview
presentations on state-of-the-art tracer systems and/or
atmospheric tracer applications. These presentations formed the
basis for discussions in five working groups dealing with dry
deposition and resuspension, global scale tracer applications, acid
precipitation, natural sources of acid precipitation and organics,
and regional air quality.
   The summaries of the discussions, the discussion group
recommendations, and the overview papers are included in this
proceedings document.

The Office of Health and Environmental Research (OHER) of the

Department of Energy (DOE) sponsored the Atmospheric Tracers Workshop

hosted by Los Alamos National Laboratory. The workshop was held in

Santa Fe, New Mexico, May 21-25, 1984. A multi-organizational steering committee guided by Dr. David Ballantine of OHER planned the workshop's content, general organization, participant identification, and objectives. The following people made significant contributions to these formative activities:

J. Heffter	National Oceanic and Atmospheric Administration (NOAA)
G. Ferber	National Oceanic and Atmospheric Administration (NOAA)
R. Engelmann	National Oceanic and Atmospheric Administration (NOAA)
P. Gudiksen	Lawrence Livermore National Laboratory (LLNL)
R. Dietz	Brookhaven National Laboratory (BNL)
R. Semonin	Illinois State Water Survey (ISWS)
T. Horst	Pacific Northwest Laboratory (PNL)
S. Barr	Los Alamos National Laboratory (Los Alamos)
W. Clements	Los Alamos National Laboratory (Los Alamos)
P. Guthals	Los Alamos National Laboratory (Los Alamos)

The general sessions were chaired by Dave Ballantine, who has also graciously provided an Executive Summary for this document. The presentations given during the general sessions are included in Section IV of this document. We want to express our collective appreciation to those who unselfishly gave of their time to prepare and give these presentations.

The discussion group chairmen contributed in large measure to the success of the workshop with their tireless efforts and deserve a special note of appreciation.

Acid Precipitation	J. Hales
Continental- and Global-Scale Tracers	B. Zak
Dry Deposition and Resuspension	T. Horst
Regional Air Quality	F. Gifford
Natural Sources of Inorganic and Organic Gases and Aerosols	J. Gaffney

All the workshop participants who took time from their busy schedules to attend played an important role in the success of this conference. The local hosts and the steering committe express their thanks to all the workshop participants. We hope that all those who made written contributions to the proceedings will recognize their work in this final document. It is our intent to publish the working group contributions as well as the formal and informal presentations as received, making only grammatical corrections.

On behalf of the workshop, we wish to acknowledge the superb support efforts by Los Alamos National Laboratory and contractor personnel, the Sheraton Hotel conference group, and the Santa Fe Chamber of Commerce volunteers. The publication of this document would not have been possible without the excellent support of Jody Heiken and Suzie Dye of Los Alamos and Pat Goulding and her staff from Los Alamos Technical Associates, Inc.

Part II

Executive Summary

EXECUTIVE SUMMARY

A workshop on atmospheric tracers was considered timely because there have been a number of significant advances during the 5 years since the previous workshop devoted to this subject. These advances include:

- Development and demonstrated use of perfluorocarbon and "heavy" methane tracer systems for dispersion studies on the scale of 1000 km.
- Development of remote sensing lidar systems for routine tracking of particles in plumes.
- Demonstrated use of specific elements and element ratios as signatures for specific regional sources.
- Demonstrated photographic techniques, both surface-based and from satellites, to measure plume dispersion.

In addition, there is increased interest in tracers and a demand for resolution of such regional- and global-scale questions as acidic deposition, visibility degradation, and climate change.

The objectives of the meeting were to

- Provide an up-to-date report on tracer technology developments and tracer applications.
- Provide a forum for scientific exchange among researchers who are active in various areas of tracer research and application.
- Examine key problem areas and identify needs for and potential applications of tracers.

The format of the meeting included an initial series of invited lectures by well-recognized scientists who provided a common base of understanding for all the participants. These lectures covered the following subjects:

- Tracers in Transport and Diffusion
- Capabilities, Needs, and Applications of Gaseous Tracers
- Measurement of Dry Deposition and Resuspension Using Tracers

- Review of Particle Tracers of Atmospheric Processes
- Atmospheric Tracers of Opportunity from Important Classes of Air Pollution Sources
- Use of Tracers for the Study of Atmospheric Chemical and Physical Transformation Processes
- Tracer Application for the Study of Precipitation Processes
- Seasonal and Diurnal Effects on Pollution Transport in the Rocky Mountain West

These presentations were followed by a series of group discussions devoted to selected "problem" areas.

- Applications of Tracers to Acid Precipitation Studies
- Continental-and Global-Scale Tracers and Their Applications
- Dry Deposition and Resuspension
- Regional Air Quality
- Natural Sources of Inorganic Gases and Aerosols

An attempt was made to include in each discussion group an appropriate mix of participants representing competence and experience in meteorology, atmospheric chemistry and physics, and numerical modeling because an understanding of atmospheric processes requires contributions from and interactions among these disciplines. Each panel was charged to analyze the status of research in its problem area, evaluate the potential contribution that tracers might make toward solutions, identify specific requirements, and develop recommendations for future study.

It was clear at the outset that there would be considerable overlap among the various discussion groups. An understanding of any one of the subjects requires an ability to evaluate the various atmospheric processes of transport and diffusion, chemical and physical transformations of gases and particles, precipitation scavenging, and dry deposition and resuspension, so it was not surprising that duplications occurred in the identification of needs and recommendations.

I. GENERAL COMMENTS

Inert tracers were considered reasonably well developed and demonstrated, so studies involving transport and diffusion on scales of 1000 km are practical. The principal inert gaseous tracers are a series of perfluorocarbons (perfluoromethylcyclohexane, perfluorodimethylclohexane, and perfluoromethylcyclopentane). At shorter distances (less than 100 km), SF_6 continues to be a widely used tracer, as are the halocarbons. The "heavy" methanes ($^{13}CD_4$ and $^{12}CD_4$) can be used on continental scales, but analytical complexity and costs currently limit their use.

The principal gaseous tracer needs are for additional compounds and more sensitive, rapid, and cheaper methods of analysis because of the large number of samples that must be analyzed in major field experiments. There is also a need for improved "real time" measurements at lower concentrations (particularly in sampling from aircraft), improved techniques for elevated releases of tracers in puff or continuous release situations, and better vertical sampling techniques.

Reactive tracers are needed to study chemical and physical transformations and wet and dry deposition. Major interest at the moment centers on sulfur and nitrogen compounds because of the interest in "acid rain." Tracers labeled isotopically with stable sulfur, nitrogen, and oxygen isotopes have been proposed. Their use in field experiments is restricted because of the limited availability of the stable isotopes and their high cost. Other limitations are imposed because of the isotope exchange, which can sometimes occur between the isotopically tagged tracers and other constituents in the atmosphere. Several panels discussed the use of radioactive ^{35}S to tag SO_2 and sulfate, but current public sensitivity to the release of radioactive materials into the atmosphere has precluded serious consideration of these tracers.

Several groups also discussed using surrogate tracers to simulate the behavior of actual chemical species in transformation, deposition, and resuspension studies. This is currently a very speculative area and before any surrogate can be used with confidence, considerable work will be required to evaluate how well the surrogate simulates the behavior of the compound it represents.

Particle and aerosol tracers were discussed by all groups, and a number of the needs defined for gaseous tracers also appeared in these discussions. Better release systems are necessary, particularly where large amounts of monodisperse aerosols are required. Generative systems for submicron aerosols and the ability to readily change the aerosol size would be particularly useful. Where polydisperse aerosol tracers are used, it would be desirable to have improved size discrimination in air- and deposition-sampling systems. Improved cloud water sampling is a specific need in the acid precipitation research community. The Deposition and Resuspension Discussion Group emphasized that definition of the chemical and physical properties of tracers used in resuspension studies must be improved if a better understanding of this process is to be developed.

Particulate tracers received considerable attention during discussions of source/receptor relationships. Tracers of opportunity have been employed in receptor modeling where specific regional signatures were identified by using ratios of elements. Ratios of V/Mn have been used on regional scales and graphitic carbon on continental scales. The use of submicron particles containing rare-earth elements was considered a promising area and worthy of high priority.

It was noted that balloons can and should be used in many field programs to help direct sampling teams to the air volumes of greatest interest. The meteorologists felt there is need for improved methods to study the dynamics of cloud systems where chaff and cloud hydrometeors are potential tracers. This need could best be satisfied by remote sensing techniques--a requirement identified by all panels.

Development of remote sensing and measurement systems was probably considered the highest priority by all discussion groups. Systems to measure specific gases and particles over large air volumes and to assess atmospheric dynamics from ground stations or airborne and space platforms would permit dramatic advancements in atmospheric studies. Doppler lidar, radar, and acoustic sounders have already made great contributions, but much more is needed. An accelerated effort in this area could pay rich dividends.

In a philosophical vein, there were discussions of the value of interactive diagnostic modeling with the planning and analysis of field tracer experiments and the value, if not necessity, of supporting tracer studies with intensive meteorological measurements. Finally, because atmospheric science encompasses the various scientific disciplines noted earlier, it is essential that continuing efforts be made to encourage and coordinate the experience and knowledge of each group in planning and interpreting tracer experiments.

II. APPLICATIONS OF TRACERS TO "ACID PRECIPITATION" STUDIES

This panel discussed the applications of atmospheric tracers to precipitation-scavenging studies and summarized their scope as being cloudscale, multicomponent, and diagnostically oriented. While recognizing that the prestorm history of an air mass could not be separated from the in-cloud processes, the panel gave it secondary emphasis. Cloudscale was seen as variable from hundreds of meters to a thousand kilometers or more. Emphasis was placed on acids and acid precursors, but the importance of all pollutant species was acknowledged. The panel specifically emphasized that the continuous iteration between field experiments and diagnostic modeling is the most fruitful approach to understanding this complex problem.

The panel felt that the precipitation scavenging sequence includes the following components:

- air flow in and around cloud and storm systems,
- attachment of pollutants to cloud droplets and hydrometeors,
- chemical transformation, and
- precipitation formation and deposition.

They pointed out that these components are interactive, with possibilities for reversibility and feedback within the sequence and that of all the processes are dependent on general air-flow characteristics. They also noted that the current understanding of each of these processes is insufficient to provide a high degree of competence in dealing with the acid precipitation issue.

The panel proposed a strategy of progressive studies, suggesting that tracer applications should begin with (1) comprehensive applications to simple precipitation systems or (2) simple applications to more complex systems, and then the applications could be extended to even more complex situations.

A hierarchy of experiments was identified, including a study of cyclonic-storm transport that is essentially the use of inert gaseous tracers to study the transport of a pollutant-laden air mass through a complex frontal system. The study would provide a data base for model evaluation and improvement. Another suggested simple experiment would use an inert tracer as a Lagrangian marker of an air mass that was expected to undergo cloud condensation; the experiment would include a comparison of the pollutant composition of the pre-cloud and the in-cloud air. A study of transformation in cloud systems that have well-defined flow fields was offered as an example of a simple complex experiment. Orographic rain systems over mountains and lake-effect precipitating systems were seen as two attractive possibilities.

The panel defined "ideal" tracers, compared these with currently available tracers, and identified tracer needs. The need for remote sensing capability was recognized for both measuring pollutants and defining flow regimes in and around clouds. As in other panels, the need for chemically reactive tracers to simulate pollutant behavior was emphasized. The special need for fast-response detectors in aircraft used to probe cloud systems was also discussed. Many needs identified here were also included in other panel discussions.

III.   CONTINENTAL-AND GLOBAL-SCALE-TRACERS AND THEIR APPLICATION

The boundaries for discussion as established by this panel were

- distances of more than 1000 km,
- duration of more than 2 days,
- altitude: surface to 50 km,
- all important processes on these scales, and
- both natural and anthropogenic phenomena.

The panel considered two broad continental- and global-scale issues that could be studied using tracer techniques: climate change and health, ecological, and environmental effects. In addition, they discussed using tracers to improve predictive capability in transport and dispersion.

The climate change discussion ranged from the effects of CO_2 and aerosol buildup to effects of nuclear war. Episodic natural events such as large volcanic eruptions were seen as opportunities for studying the potential effects of anthropogenically initiated climate perturbation.

In the area of health, ecological, and environmental effects, the discussion ranged from international transport of toxic materials and acid deposition precursors through global-scale visibility reduction to the impact on the ozone by persistent gaseous species.

Concerning the low level of predictive capability, the discussions dealt with the limitations on information regarding global budgets, source distributions, transport, transformation and removal processes, and the problem of defining appropriate boundary conditions.

Use of controlled-tracers on continental and global scales would require either an increase in the quantities released or improvements in the sensitivity of measurement. Currently, only selected perfluorocarbons and the heavy isotope analogues of methane ($^{13}CD_4$ and $^{12}CD_4$) have been demonstrated beyond the 1000-km range. Most of the discussion centered on natural tracers and anthropogenic sources of opportunity. The natural tracers included (1) intercontinental transport of soil dust, organic particles, volcanic gases, and aerosols for stratospheric and tropospheric global transport and stratosphere-troposphere exchange, (2) ^{222}Rn for land-to-ocean transport, and (3) ozone for stratosphere-troposphere interactions. The anthropogenic tracers that were discussed included trace element ratios, halocarbons, and graphitic carbon. The panel also considered the use of controllable, constant-volume balloons that could be tracked by satellite.

IV. DRY DEPOSITION AND RESUSPENSION

A. Deposition

This panel chose to address deposition in terms of (1) particles, (2) gases, and (3) generic issues.

For particles, attention centered on:

- the particle-size dependence of deposition velocity in vegetative canopies, particularly deeper canopies, to resolve the differences between theoretical and experimental results;
- the adequacy of sampling methods for supermicron particles in polydisperse particle-size distributions;

- the modification of particle sizes, and therefore dry deposition, caused by water-vapor condensational growth of particles; and
- the bounce-off of supermicron particles from natural surfaces.

The questions related to gases included:

- surface supersaturation/nonlinear effects on dry deposition,
- the influence of co-pollutants on dry deposition of selected species,
- the dry deposition of hydrocarbons and free radicals, and
- transfer velocities in surface media.

Generic issues included:

- uncertainties in total deposition that result from summing the deposition of various component surfaces;
- the resolution of uncertainties about existing dry deposition monitoring techniques;
- the measurement of dry deposition in urban areas; and the adequacy of chamber and wind tunnel studies for simulating field measurements.

The panel felt that studying dry deposition of particles should involve both field and laboratory work. Existing tracers were considered adequate for laboratory studies. For field studies, however, there is need for both an improved system to release large amounts of monodisperse aerosols and an improved method to examine air and deposition samples (that is, a technique capable of determining the number of particles in limited size ranges). The use of dual tracers (depositing and nondepositing) for measuring total deposition over large distances was discussed, but successful application awaits improvements in vertical sampling. Remote sensing capability was identified as a high-priority need.

The tracing of gases appears to be done best by using stable and/or radioactive isotopes of the gas of interest. Current limitations on the availability of the stable isotopes of interest and perceived public reaction against the use of radioactive isotopes preclude their use in deposition studies at this time.

B. Suspension and Resuspension

The panel underscored the general importance and broad range of problem areas associated with resuspension, which ranged from the emission from various types of industrial and municipal dumping and storage areas to concern over soil erosion. There was a strong feeling that research in this area is hampered by inadequate funding and lack of a fundamental understanding of the chemical and physical processes that control suspension and resuspension.

The panel did not feel that tracer technology is limiting research progress in this area. However, there are clear indications that more studies are needed to define the physical and chemical properties of tracers that are used. In particular, there is need for better characterization of tracer materials and their behavior so that the chosen tracer is a more realistic surrogate of the original material.

Two examples were given of problems in which tracers could be used to understand physical processes: (1) locations where large amounts of toxic/hazardous materials have been deposited and may be entrained into the atmosphere and transported downwind and (2) the general "erosion" of particulate material, which can have long-term effects at the site or downwind.

V. REGIONAL AIR QUALITY

The regional scale adopted by this panel extends loosely from local scales of 10s of kilometers to continental scales of 1000s of kilometers; this spectrum allows for the concept of air quality regions and the mesoscale of meteorological phenomena. The phenomena included:

- urban-rural transport and dispersion,
- complex terrain effects,
- coastal effects,
- cloud processes and organized convection system effects, and
- dispersion of hazardous materials.

A. Urban-Rural Transport and Dispersion

The panel noted that verifying source/receptor relationships at regional-scale distances is confounded by difficulties in defining the transport and diffusion and by the various processes of transformation and deposition that can occur during transport. A number of these considerations are treated in detail in the discussions of the Deposition and Acid Precipitation Panel. Transport and diffusion on the regional scale can now be handled rather well by conservative perflourocarbon tracers. There are still needs for additional perfluorocarbon tracers, for more versatile tracer release systems so that elevated releases of up to 1 km (either puff or continuous) can be carried out, improved vertical sampling capability (again to 1 km), for real-time measurements at lower concentrations, and ultimately, for sensitive remote sensing capability.

Nonconservative tracers (water soluble, chemically reactive, and depositing tracers) were recognized as key needs for regional-scale studies. The panel felt that rare-earth elements, particularly in submicron-scale aerosols, show considerable promise as depositing tracers and were a high-priority research area. Other nonconservative tracers are not as well developed. In all cases, there is a need to evaluate critically how well the tracers simulate the emissions of interest.

Inherent tracers for specific types of sources (coal-burning power plants, smelters) or regions (Midwest, New England) are feasible for identifying source-receptor relationships. These are usually aerosol particles containing inorganic signatures. It was suggested that consideration be given to looking at the utility of organic compounds or inorganic vapors.

B. Complex-Terrain and Coastal Transport

Complex terrain and coastal transport are really subsets or subcomponents of the regional-scale problem, and the tracer needs were quite similar to those identified under urban-rural transport and dispersion. A general caveat for future studies was the necessity of determining the representativeness of release and sampling sites in complex terrain areas. In coastal studies, reliable sampling systems must be developed for use at the surface layer over water.

The discussion of cloud processes and organized connective systems effects duplicated that in the Acid Precipitation Panel, with similar

conclusions and identified needs. Similarly, the discussion of hazardous
material dispersion was a repetition of some material covered by the
Deposition and Resuspension Panel.

VI. NATURAL SOURCES OF INORGANIC AND ORGANIC GASES AND AEROSOLS

It is clear that both natural and anthropogenic sources of organics
(including sulfur-, nitrogen-, and halogen-containing compounds), oxides of
nitrogen and ammonia, and sulfur compounds in gas and aerosol phases can
contribute to the observed composition of acid precipitation, gaseous and
aqueous oxidants, degradation of visibility, cooling and/or heating of the
troposphere, and other atmospheric effects.

The two source types differ in that generally natural sources are
distributed, whereas anthropogenic sources are largely point sources.

The working group focused on the use of tracers to differentiate natural
from anthropogenic sources of inorganics, organics, oxides of nitrogen,
ammonia, and sulfur compounds in gaseous and aerosol form. The natural
sources of these materials included:

- emissions from vegetation, soils, and biota (land),
- forest fires (including wood and agricultural burning),
- soil erosion and suspension,
- marine sources,
- volcanic activity,
- atmospheric electrical activity, and
- stratospheric injection.

In each category, the panel identified specific compounds that could
possibly be used as signatures for that natural source. Because in some cases
the natural primary emission may have a limited lifetime in the atmosphere,
the need to consider stable reaction products was emphasized. Because
$^{13}C/^{14}C$ ratios and ^{14}C measurements are useful tracers for many natural
sources, the panel emphasized the desirability for improved ^{14}C analysis.

The panel emphasized the need for interactions among atmospheric scientists, chemists, geochemists, and plant and soil scientists to establish the best time and locations for evaluating natural sources. The panel also noted that the spatial and temporal variability of natural sources will require source variability studies as a function of humidity, region, and season.

The use of advanced statistical analytical methods to evaluate multicomponent source terms was also recommended.

Part III

Discussion Group Summaries

APPLICATION OF TRACERS TO "ACID PRECIPITATION" STUDIES

I.   INTRODUCTION: DEFINITION OF SCOPE

The objective of this working-group report is to discuss applications of atmospheric tracers for precipitation-scavenging studies. Several other working groups are considering closely related subjects, and it is important at the outset to define the boundaries of this report. These are established primarily upon the basis of scientific considerations as well as known discussion areas of the other working groups, and are described below.

- The physical domain of our primary interest is the real atmosphere and what we loosely describe as the "cloud scale." Transport, deposition, and physiochemical reaction before a pollutant's encounter with a cloud are noted as strong determinants of in-cloud behavior and thus cannot be completely separated from this discussion. Regardless of that fact, prestorm history will be given a somewhat secondary emphasis in this report. It is also noted that, depending upon storm features, the "cloud scale" can range from a few hundreds of meters to 1,000 kilometers or more. Finally, it is important to note that although this report deals exclusively with atmospheric tracer applications, several potentially important advances in this general area are possible with laboratory application as well.

- All pollutant species will be considered important throughout this discussion; however somewhat higher emphasis will be placed upon acids and acid precursors. The primary reason for this generalized approach is that because of the commonality of many elements of the scavenging process, several of the tracer applications will have large universal utility.

- It is noted that two general categories of tracer application are possible, which we shall term here as "empirical" and "diagnostic." The empirical approach is best represented by the MATEX concept, where massive quantities of tracers are released with

20

the hope that the accompanying measurements will lead to obvious empirical relationships describing source attribution and other important features. The diagnostic approach involves application of tracer in conjunction with some sort of deterministic model--either to supply necessary input information to a prognostic model or to enable a diagnostic model to elucidate mechanistic behavior. There have been recent comprehensive reviews of empirical tracer applications, and for this reason this report will place prime emphasis on applications of the "diagnostic" type. In doing this, however, it is noted that many potentially important empirical applications are possible and may give rise to significant advances in our understanding of wet-removal processes.

The scope of this discussion can be summarized as being cloud-scale, multicomponent, and diagnostically oriented. In the following text, we shall attack this subject by first identifying those of the precipitation-scavenging processes that (1) are currently limiting our capability to describe and predict wet-removal behavior, and (2) can be (in principle) elucidated by tracer techniques. We shall describe then several candidate experiments where such measurements can be made and attempt to identify both strong and weak aspects of the conceptual designs.

The next section describes the physical and chemical characteristics of the "ideal tracers" required for the previously defined experiment designs. The report concludes with a discussion of the advances in tracer technology that are needed to upgrade our present capability to levels required for these more advanced applications.

II. KEY SCAVENGING PROCESSES

The primary goal of this subsection is to summarize those pertinent phenomena within the precipitation-scavenging sequence that are amenable to measurement by tracer techniques. The emphasis at this point will be to segregate components of this sequence so that they can be examined in individual detail. Later in the discussion, we shall recombine many of these features--especially when composite field experiments are considered.

A rough breakdown of the components of the scavenging sequence is given below.

- Air flow in and around cloud and storm systems
- Attachment of pollutants to cloud droplets and hydrometeors
- Chemical transformation
- Precipitation formation and deposition

These components can be compared with the sequence flow chart in Fig. 1. Important aspects to note from this schematic are the multiple interactive pathways of the scavenging process, the possibility for reversibility and feedback within the sequence, and the dependency of all the processes on general airflow characteristics. It is important also to note that our current understanding of each of these processes is insufficient to provide a high degree of competence in dealing with the acid-precipitation issue. Furthermore, ideal tracers, if available, would provide an immensely valuable tool for elucidating each of these items and improving this competence.

General questions related to tracer application in this regard are itemized in Table 1. In the subsection immediately following, we shall describe a number of candidate field-study designs, that are intended to fulfill current needs within these application areas.

III. CANDIDATE FIELD STUDIES USING TRACER TECHNOLOGY

A.   Introduction

In evaluating candidate field studies the Committee established a number of basic principles, or "design criteria," which are itemized below.

- To achieve success, a tracer study must be accompanied by a substantial ensemble of supporting measurements. In general,

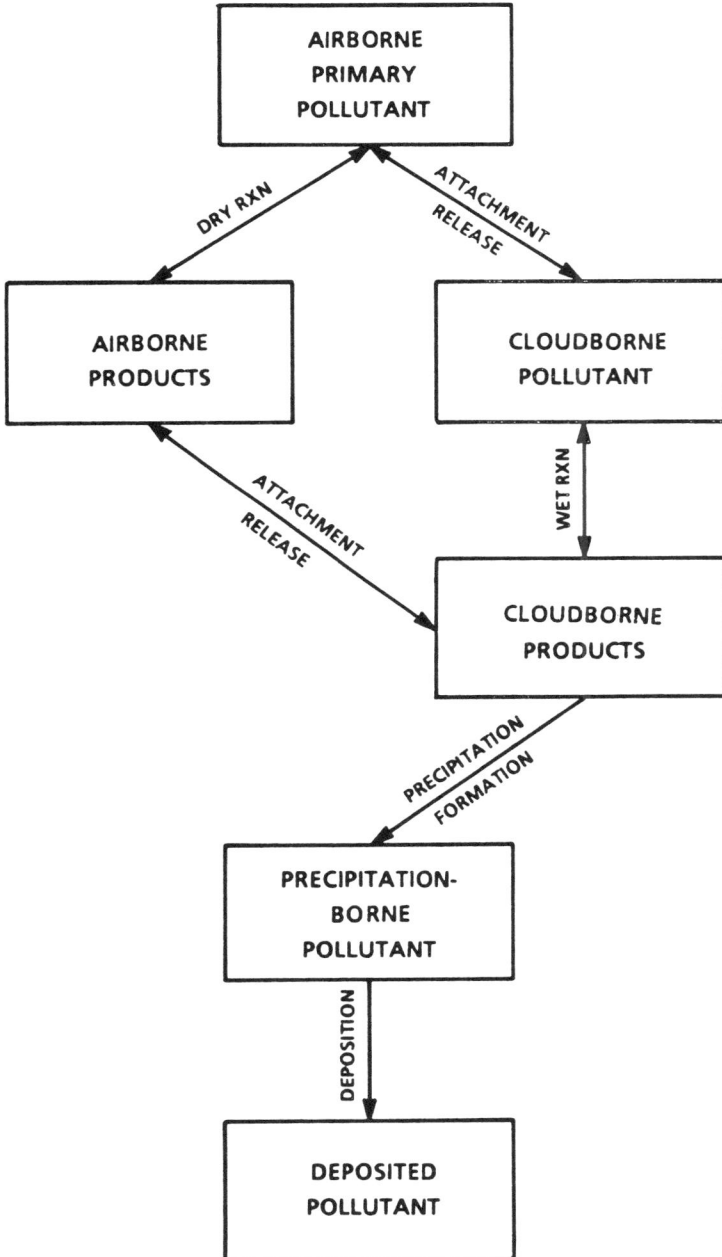

Fig. 1.   Simplified schematic of scavenging sequence.

TABLE I

GENERAL QUESTIONS TO BE ADDRESSED BY
TRACER APPLICATIONS AS RELATED TO KEY SCAVENGING PROCESSES

- What is the characteristic nature of the 3-dimensional field of air motions in and around various types of clouds and storm systems that determines if, when, and where a pollutant of interest will enter the systems?

- After a pollutant has entered into precipitating or nonprecipitating clouds in a weather system, what will be its 3-dimensional trajectory? Where will it eventually be deposited on the ground in precipitation? How is this influenced by type of weather system, pollutant release location relative to the weather system, and time of year?

- How many cloud-entrainment cloud-detrainment cycles does a pollutant typically undergo before it finally is removed by precipitation scavenging, and how does this affect the final chemical form of the pollutant material and the resulting acidity of the precipitation?

- Given that a pollutant of interest enters a cloud or storm system, how much of the pollutant is removed by the precipitation scavenging process and deposited on the ground, how much passes through, and how much of it is left as a residual when the cloud dissipates? How do these amounts depend upon various factors, such as

  - type of pollutant,
  - type of cloud,
  - type of weather system,
  - time of day,
  - season, and
  - presence of other pollutants or natural species.

- What chemical transformations occur within clouds that influence precipitation acidity, and how are these affected by pollutant mix and ambient conditions?

---

tracer application should be considered only as one component of a multifaceted field study.

- There often will be a synergistic benefit to the concurrent release of multiple tracers, each having different response characteristics within the scavenging system. Future field-study design should attempt to pool multilaboratory efforts to enhance this possibility.

- The systems of study tend to be extremely complex, and for this reason we must continue to think of tracer studies in a progressive sense. Tracer applications should begin with (1) comprehensive applications to simple precipitation systems or (2) simple applications to more complex systems. These can be extended outward to more complex situations as our understanding progresses.

- There is a strong need for tracers that can characterize behavior in three dimensions. Such techniques should be encouraged whenever possible in preference to those requiring point measurements.

- Relative geometries of the tracer's distribution and the pollutant plume's distribution are extremely important considerations for both logistical measurements and data interpretation. These factors must be considered seriously in field-experiment design.

- The use of heterogeneous aerosol tracers should be discouraged, except for purposes of qualitative (or at best, semiquantitative) target-marking activities. The development of homogeneous aerosol tracers should be strongly encouraged.

B.   A Hierarchy of Field Experiments

In this subsection we shall suggest three candidate field experiments that generally reflect the precepts given above. These are arranged in something of a hierarchical order; tracer applications are intended to probe scavenging processes in more and more detail as the progression continues. Additional types of field experiments are considered briefly in the final sections of this report.

1. Cyclonic-Storm Transport Measurements. This experiment deals only indirectly with scavenging processes, and is based upon our current strong need for a better understanding of pollutant transport through complex frontal systems. At present, we have low confidence in our ability to predict,

describe, or otherwise simulate the behavior of even an inert pollutant in such circumstances, and until this competence is achieved, our capability to model and understand acid precipitation will remain unacceptably low.

This experiment can be test described as "CAPTEX-enhanced." It involves the CAPTEX-like release of a single inert tracer into a passing frontal system, with arrays of airborne and surface sampling equipment located downwind. Such an experiment must be accompanied by comprehensive wind-measurement facilities.

Subsequent to the physical experiments, a variety of techniques should be tested to establish our proficiency in simulating pollutant transport in frontal systems. If this proficiency is acceptable, it will be expedient to progress in acid-deposition model development and more comprehensive experiment design. If not, a fundamental area for future research emphasis has been revealed. The Committee feels that this experiment is an important progressive step in our state of understanding.

2.   Tracers as Short-Range Air-Parcel Markers. In contrast to the previous experiment, this simple candidate field study involves direct measurement of pollutant capture processes. It is a potentially valuable experiment that can be conducted with present capabilities and can be viewed as a "simple tracer simple storm situation." Such an experiment deploys a tracer as a Lagrangian marker to identify (tag) a parcel that is expected to soon undergo cloud condensation. Changes in composition between pre-cloud air and in-cloud air are then unambiguously ascribed to cloud formation.

3.   A Simple Comprehensive Experiment. The most straightforward way to use tracers for elucidating in-cloud chemical processes involves cloud systems that have well-defined flow fields and permit logistically tractable operations. Two attractive possibilities exist: orographic rain systems over mountain ranges and lake-effect precipitating systems. Both have the advantage of being (more or less) fixed in space and possess relatively steady-state flow fields--at least over periods of a few hours. Our "simple comprehensive" experiment involves tracers used in the following way:

- chaff or dyes tracked by remote sensing to determine flow fields,
- tagged SO_2,
- tagged NO_2,
- two conservative tracers (for example, perfluorocarbons), perhaps with one ramped to give time resolution,
- highly soluable gaseous tracer, and
- aerosol tracer (uniform size).

In situ monitoring would be done by aircraft using chaff as a marker to determine where to fly to pick up other tracers. An extensive ground network for sequential precipitation sampling would be required.

Examination of the extent of chemical processing by the cloud and delivery to the surface in precipitation would be made from ratios of concentrations of tagged sulfur to inert tracer. This ratio of tagged SO_2 to inert gas in the outflow region of the cloud would yield an unambiguous determination of the amount of SO_2 removed by cloud. (It would be valuable as well to ascertain the amount of tagged SO_4 present in this outflow region.) Similarly, the ratio of tagged SO_4/soluable gas in precipitation samples would establish the fraction of SO_2 that we can assume dissolved and reacted immediately in the cloud (for example, with H_2O_2). Tagged aerosol particles [for example, low mole-fraction rare earth in a matrix of $(NH_4)_2SO_4$] could be examined in the same way: ratio to inert gas in transmitted air, ratio to soluable gas in precipitation samples.

In such an experiment, it will be extremely important to have a full complement of chemical support measurements, including concentrations of inflow acids, acid precursors, oxidants, and components involved in the chemical oxidation process. These later components include (but are not necessarily limited to) hydrocarbons and other organic constituents and oxides of nitrogen.

IV. IDEAL TRACERS AND DEVELOPMENT NEEDS

Tracers used in the atmospheric sciences are chosen on the basis of chemical and physical characteristics that are important in selected processes. The materials chosen may be subject to one, several, or all of the processes that control the behavior of an atmospheric constituent or class of constituents. In precipitation-scavenging studies, a wide range of tracer categories is useful. These categories include both reactive and inert gases as well as aerosols and can involve artificially inserted tracers as well as existing "tracers of opportunity." Table II summarizes characteristics of ideal tracers and indicates some of the limitations that occur in reality. Items 4 and 7 imply an assumption that tracer-induced perturbations to the "natural" environment and limited spatial sampling are always undesirable. However, some experimental designs may call for intentional, well-controlled perturbations or spatially selective probing with tracers. If not, tracers of opportunity have the advantage. In addition, study of the tracers themselves may lead to a better understanding of scavenging and oxidation processes.

The itemized discussion that follows concentrates first on the characteristics of artificial tracers. We assume that substances can be released and sampled where needed. In many cases, a combination of aircraft deployment and surface sampling networks is required.

A. Inert Tracers

These are nonreactive, which means that they do not change chemical form. They are used to tag parcels of air through complicated patterns of flow and diffusion. For the lifetime of the clouds or storm systems studied, inert tracers must remain unaltered, not be scavenged by rain and not become attached to ambient particles or cloud droplets (unless intentional). Inert tracers can be classified according to whether they are used in conjunction with remote sensing techniques or whether they require in situ sampling.

1. Remote Sensing. Particles are the most commonly used materials for remote sensing applications. For acid precipitation studies, tracking the material within clouds is extremely desirable. A prime example is detection of chaff by radar. Ideally, the tracer has a negligible gravitational settling

TABLE II

SUMMARY OF TRACER CHARACTERISTICS FOR
ACID PRECIPITATION STUDIES

IDEAL TRACERS

Artifical Tracers

Zero background concentration

Sufficient quantities available in proper chemical and physical form

Efficient release methods

No perturbation to atmospheric physical and chemical mechanisms

Artificial Tracers and Tracers of Opportunity

Excellent analytical accuracy and precision

Efficient collection or detection methods

Precondensation distribution in air mass is identical to that of pollutant materials studied

Known cloud processing and scavenging properties

REALITY

Artifical Tracers

Low background, not always negligible

Isotopes of low-abundance that are materials that are often costly or difficult to handle

Release techniques often a major difficulty in amounts of materials released or in placement of releases near clouds

Care needed not to release quantities too large

Artificial Tracers and Tracers of Opportunity

Adequate accuracy and sensitivity often achieved with time-consuming and costly analysis

Poor time resolution for some in situ collection procedures; costly remote sensing systems

Poor spatial coverage with artifical tracers

Except for inert or very reactive gaseous tracers, inadequate knowledge of cloud processing and scavenging processes velocity to provide a true depiction of updrafts and downdrafts and does not collect droplets to a significant degree (this might increase settling velocity and even disturb cloud microphysics and dynamics).

An alternative approach is to use the hydrometeors or droplets themselves, as viewed by radar, to obtain information on flow regimes within clouds. To measure wind velocity fields near clouds, other techniques can be used also; sensing of refractivity fluctuations or aerosol concentrations becomes feasible.

In some cases, existing analytical capabilities are not being used to their full potential. Routine monitoring frequency should be increased, in conjunction with increased radiosonde soundings for the characterization of mean advection. Remote sensing techniques (for example, Doppler RADAR and airborne LIDAR) for flow and atmospheric structure in three dimensions could be used more effectively if they were routinely available and more convenient to apply.

2. In Situ Sampling. Tagging air parcels with inert gaseous tracers such as SF_6 or perfluorocarbons is sometimes applied in conjunction with measurements of ambient conditions. The purpose is to identify air parcels along the course of travel through clouds. Many of the requirements have already been mentioned. It is desirable to have a suite of inert gaseous tracers so that several experiments involving multiple trajectories can be performed almost simultaneously within the same cloud or storm system.

Just as hydrometeors can be used in remote sensing, the in situ sampling of rainwater can be used as a chemical tag, by examining the isotopic composition of oxygen and hydrogen. The isotopic compositions of oxygen and hydrogen have not been used to validate models of atmospheric mixing and other processes. Applications that could be developed include identifying the degree of mixing of two air masses, the distribution of water between vapor and liquid phases in clouds, and the amount of water removed by precipitation.

B. Particle Tracers

Often referred to as "aerosol tracers," fine particulate materials have been used as tracers that are released in cloud inflow regions and later collected in rain samples from a surface network of precipitation collectors. The particulate material does not serve as an inert tracer because it is scavenged, but it must remain in a form easily identifiable. Examples include rare-earth elements released separately or in doped sulfate aerosols.

The use of a "monodisperse" aerosol is desirable to study attachment and scavenging processes. The development of improved aerosol generators that are capable of releasing cheaply labeled submicron particles in a very narrow size range and in sufficiently large amounts remains to be accomplished. A multitude of cheap particle tracers are needed to tag different air masses, and the ability to vary the particle size of these monodisperse aerosols is also desirable.

Further, the ability of released particles to act as cloud condensation nuclei and to accrete water are processes that should be understood. A lack of such knowledge can limit the usefulness of particle tracers in clouds, even if the particles can be released as a monodisperse aerosol.

C.   Chemically Reactive Tracers

Another class of ideal tracer includes gases that are irreversibly soluble in cloudwater and rainwater. Ideally, this gas should not partition with ambient particulate forms and thus would remain in a unique gaseous form until scavenged irreversibly by cloud droplets or raindrops. Such a gas would provide an excellent surrogate for ascertaining the uptake by clouds and deposition in precipitation of nitric acid gas. The tracer would also set an upper bound for the gas dissolution of sulfur dioxide, which requires additional chemical oxidation reactions to attain any irreversible dissolution. Ratioing of tagged sulfate to the irreversibly soluble tracer would allow the determination of the extent of these chemical reaction limitations.

The limitations on the development of an irreversibly soluble gas are twofold. First, the gas should not perturb the acid-base, oxidation-reduction, or mass-transport properties of the scavenging processes. Second, it is difficult to identify a chemical that is volatile, infinitely or irreversibly soluble in cloud water and rainwater, and also insoluble in aqueous aerosols. Trifluoroacetic acid has been proposed as a surrogate for nitric acid gas.

D.   Reactive Isotopes

Sulfur isotopes in SO_2 would be especially useful for studying processes of oxidation and scavenging. The use of a suite of sulfur isotopes is envisioned for some experiments. The isotopes ^{34}S and ^{36}S are the most often considered, although ^{35}S has the practical advantage of easier quantitative analysis, and smaller amounts are needed because the background concentrations are so low. However, it is difficult to obtain permission to use this unstable isotope in the United States. Alternate stable isotopes should be considered that, in addition to ^{36}S, might be preferable to ^{34}S because smaller amounts can be used to achieve concentrations significantly above the naturally occurring amounts. Measurements of the isotopic composition of oxygen in sulfate can also provide information concerning the mechanism of sulfate formation.

Although isotopic labels of sulfur and oxygen have been used for sulfur dioxide, the analogous work with nitrogen oxides has not been (but should be) considered for validating the relative importances of different mechanisms for NO_X scavenging. The development of field studies using ^{15}N- and ^{18}O-labeled nitric oxide and nitrogen dioxide appear useful. It would be necessary to confirm that oxygen exchange rates are slow before considering oxygen isotope labels for nitrogen oxide species.

Whereas isotopic tracers have the potential for being nearly ideal tracers for these reactive species, the cost of these tracers and their analysis appear to be the limiting factors. Options for the development of isotopic tracers include (a) decreased costs if a large-scale capacity for isotope production and is developed, (b) increased sensitivity of analysis, or (c) identification of the maximum scale of precipitation event that can be studied with available techniques.

E. Tracers of Opportunity

Elements already present from natural and anthropogenic contributions offer the great advantage of being distributed in the atmosphere in a more realistic manner than locally released artificial tracers. One property of a useful tracer is a partitioning between gaseous and particulate forms that is different than the partitioning of the elements or compound being studied. Then, by examining concentration ratios of selected elements to sulfates in cloudwater and rainwater, for example, the direct contributions of ambient sulfate aerosol to sulfates in rainwater might be determined. This area of research currently needs more work to evaluate the natural tracers that might be useful.

It has been pointed out that in areas distant from pollutant source regions, the ratio of SO_4^{2-} to Se in clear-air aerosol particles seem remarkably uniform. It may thus be reasonably expected that significant in-cloud formation of SO_4^{2-} from SO_2 would substantially alter this ratio in cloud or rainwater. It appears that the feasibility of measuring SO_4^{2-}/Se ratios in clear air and precipitation as a measure of in-cloud SO_2 oxidation could be readily established in a limited number of diagnostic experiments.

Within the tracers of opportunity, several needed general developmental tasks have been identified.

·   Analytical capabilities must be improved to lower the minimum mass detection limit, or sampling volume must be increased to provide sufficient mass for accurate quantification in the desired time resolution within a precipitation event. This applies to: organic constituents, stable isotope ratio determinations, and trace metal species analyzed by atomic absorption, neutron activation analysis, or x-ray fluorescence. Improvements in in-cloud sampling to capture higher volumes of cloud water and interstitial air are needed.

·   Natural radioactive tracers of boundary layer (Th/Rn daughters) and stratospheric (Be) contributions are not being used to their full capabilities. The radioactive measurement capabilities should be evaluated and rain sampling handling times minimized to allow the quantification of isotopes with short half-lives. New techniques for analysis in the field may be necessary.

- Techniques to distinguish trace metals from fossil fuel emissions from crustal components should be improved, and the available techniques should be applied more rigorously to precipitation samples.

F. Nonideal Qualitative Tracers

The sophistication and costs of analysis for the above proposed nearly ideal tracers make it desirable to analyze a minimum number of samples. The use of nonideal tracers in conjunction with the more sophisticated tracers can indicate which air or water samples should be analyzed for trace species that act more nearly ideally. For example, a "smart" balloon could be developed to replace tetroons in tracking air parcels. A cheap microprocessor to sense temperature, pressure, and relative humidity would enable the balloon to follow an isentropic trajectory. With a transponder, the balloon would continuously track air parcels that had been tagged with the nonapparent chemical tracers. Thus, further improvements in nonideal tracers, without being costly, could be advantageous. Because of radon motion in the atmosphere, several (perhaps 10s) of balloons would have to be released in some cases to track an "air parcel" effectively. These nonideal tracers could give a qualitative analysis and reduce the quantitative sample analysis required for more ideal tracers.

CONTINENTAL- AND GLOBAL-SCALE TRACERS AND THEIR APPLICATIONS

I. INTRODUCTION

The Working Group on Continental- and Global-Scale Tracers felt that their topic had received too little attention in the general session. Consequently, their deliberations began with relevant presentations by J. M. Prospero, K. A. Rahn, A. Turkevich, R. Malone, J. L. Heffter, and B. D. Zak. Summaries of four of these presentations are in Section IV of this report.

After the presentations, the working group moved on to define their topic area and to outline the discussion which follows. Continental and global scale was taken to mean

- distances of more than 1000 km,
- durations of more than 2 days,
- altitude: surface to 50 km,
- all important processes on these scales, and
- both natural and anthropogenic phenomena.

II. ISSUES

Continental- and global-scale issues that are candidates for study using tracer techniques to fall within three broad categories: climate change; health, ecological, and environmental effects; and improvement of the low level of existing predictive capability (impact assessment).

A. Climate Change

Climate change could have catastrophic effects on mankind. Minor changes in climate can have major impacts on food production and other essential activities. There are now several plausible means by which man may inadvertently bring about such change. Yet even the natural causes of changes that occurred in the past are still unknown.

Man's activities have significantly increased the atmospheric load of myriad gases and particulates with diverse physical and chemical characteristics. These gases and aerosols have the capability of changing the distribution of water and ice clouds, precipitation, and the reflection and absorption of solar and thermal radiation by clear air and clouds. There are some documented examples of inadvertent weather modification, but most such modifications have occurred gradually and without pre-existing baseline measurements. Similarly, a number of anthropogenic gases that are emitted in large quantities have radioactive properties that suggest they could have a significant impact on climate. There is good reason to believe that man could be changing weather on continental and global scales, but these changes have not generally been quantified.

We can site a number of specific climate-related issues: effects of the gradual addition of radiatively active particles and gases to the atmosphere through urban and industrial activities (CO_2, smoke, etc.), the possibility that particles produced in a nuclear war might result in a precipitous drop in global temperature for an extended period ("nuclear winter"), the possible impacts of natural cataclysmic events (volcanic eruptions, meteor impacts, etc.). Each of these continental- and global-scale issues involve a host of technical questions, many of which can best be investigated using tracer techniques.

B. Health, Ecological, and Environmental Effects

Large-scale health, ecological, and environmental effects of various kinds are also plausible in light of the experience over the last few decades. For example, much of the world has suffered a marked decline in visual air quality, not just in and near urban areas, but in remote areas such as the arctic as well. In view of the fact that over 30,000 chemicals are manufactured in the U.S. alone, the potential for severe large-scale problems is considerable.

The issues of concern range from international transport of toxic materials and acid deposition precursors by the atmosphere, through global-scale visibility reduction, to impacts on the ozone layer by persistent gaseous species.

C. Low Level of Predictive Capability

The final and broadest issue, the low level of predictive capability, concerns the fact that on continental and global scales, our ability to accurately model the relationships between causes and effects of all kinds is very poor. Human activities are affecting the global environment. New activities are being pursued and the level of existing activities have drastically changed in the absence of reliable information on the consequences of these actions. Society needs the ability to accurately anticipate the consequences of its actions. The catalogue of ignorance of concern is long. It includes:

- appropriate boundary conditions to use for regional modeling;
- global budgets for gaseous and particulate species known or suspected to be important to atmospheric processes;
- spatial and temporal distribution of sources and sinks for atmospheric contaminants;
- dominant pathways for injection, transport, and removal of contaminants to the atmosphere;
- mechanisms by which the chemistry of the atmosphere could be significantly altered; and
- global circulation patterns and how they are affected by anthopogenic activities.

This list is by no means exhaustive, but it does give one a sense of how much work must be done before enough will be known for mankind to intelligently manage the global environment.

III. SUITABLE TRACER SYSTEMS

Tracers could be used to investigate many of the critical processes cited above. For such purposes, the characteristics of an ideal tracer are listed below.

- It is unique and unambiguously identifiable.
- It is measurable at extremely low concentrations.
- The material is cheap to manufacture.
- Measurement techniques are simple and inexpensive.
- The chemical and physical properties of the tracer mimic those of the chemical species of interest.
- For tracers of opportunity, the characteristics of the source can be determined.

Ideally, we should have at our disposal a variety of tracers that would enable us to characterize all atmospheric transport, transformation, and removal processes. The simplest and most tractable approach to understanding transport processes in isolation is to use relatively inert tracers. This approach avoids the complications of chemistry changes during transport. Unreactive gases with long lifetimes in the atmosphere are relatively unaffected by scavenging; these can be used as conservative tracers. Many particle tracers are relatively unreactive chemically but are subject to scavenging processes; these could be used to develop our understanding of transport and removal processes. Eventually, it should be possible to make tracers that have chemical and physical properties similar to those of important atmospheric species.

Because of the large spatial and time scales inherent in long-range transport, considerable mixing and dilution takes place during transit. This places great demands on any proposed experiments using man-made tracers (that is, intentional tracers). Currently available tracers and the associated collection and analysis techniques permit great sensitivity. Nonetheless, for the scales under discussion, their utility ranges from marginal to grossly inadequate. Large quantities of material would be required for such releases.

It is probable that cheaper tracers will be developed, that detection sensitivity will be improved, and that eventually, global-scale man-made tracer experiments may be possible.

In the meantime, there are a number of options available for these studies. Some of them make use of chemical species that are emitted in large quantities from natural or anthropogenic sources. We refer to these as "tracers of opportunity."

A. Natural Tracers of Opportunity

Natural processes emit a large number of materials in huge quantities. Because many of these emissions are relatively site specific, they can serve as tracers.

1. Soil Dust. About one-third of the continental land mass is arid, and dust storms occur in many of these regions. The sources of dust storms can be relatively well defined. They occur under specific meteorological conditions and can be observed by satellite. The soil dust size distribution attains a mass median aerodynamic diameter of about 2 μm at a distance of about 1000 km from the source. Dust clouds have been followed for over 10,000 km. Networks of air sampling stations in ocean regions have shown that dust storms can affect immense areas. Meteorological studies currently in progress suggest that the dust raised in these storms can serve as an excellent tracer. Network measurements are being used to develop trajectory techniques and to validate global circulation models. Because certain areas (for example, in North Africa and Asia) are known as frequent and predictable sources of dust storms, it should be possible to design specific tracer field experiments.

2. Organic Particles. A major portion of the particulate organic material over the oceans is derived from continental sources. The composition of this organic material is largely unknown. However, plant leaf waxes have been identified as an important class of compounds. Plant wax particles are relatively inert, and they are predominantly in the submicron

size range. The composition of these waxes is often specific for certain classes of plants. Consequently, it may be possible to identify the source of a specific air parcel on the basis of the plant wax composition.

Pollen and spores have also been used as tracers on this scale. More is said about these in Gatz's presentation on particulate tracers (Section III).

3. Radon-222 and Daughter. Radon-222 is a radioactive inert gas (half-life = 3.8(?) days) that is emitted predominantly from soils. The rate of emission of ^{222}Rn from the ocean is at least 100 times less than that from the continents. Thus, ^{222}Rn and its daughters are non-site-specific tracers of continental air parcels. Radon-daughter products immediately become attached to aerosol particles. By measuring the concentration of radon and its daughters and the concentration of the daughters in precipitation, it should be possible to investigate the global transport of surface-emitted materials from the continents to the oceans.

4. Volcanoes. Volcanoes are prolific, if highly sporadic, producers of particles and gases. They generally introduce material into the middle and upper troposphere, but they often inject material into the stratosphere and are a major source of sulfur for the stratospheric sulfate layer.

5. Ozone as a Tracer. Because there is an international effort to measure the ozone distribution, ozone offers a valuable opportunity for verifying the mixing calculated by global circulation models and for increasing our understanding of stratosphere-troposphere exchange processes. The advantages of ozone as a tracer are that its stratospheric source is rather well known and that it can be measured with sufficient accuracy (for use as a tracer) with existing equipment. Smaller sources exist within the troposhere. It is presently measured by surface stations, balloon-borne sondes, and occasionally, by instrumented aircraft. The data for this international effort are collected in the center in Canada. Calibration and intercomparisons are handled by WMO. An enlarged ozone-sonde effort to complement the existing network observations could provide a continuing data base for estimating stratospheric transport and mixing.

B. Man-Made Tracers of Opportunity

One of the major motivations for carying out tracer experiments is to assess the extent and degree of man's impact on nature. If the anthropogenic product is different than that emitted from natural sources, the product can serve as its own tracer. Anthropogenic tracers are especially attractive from the global standpoint because there are no oceanic sources for many species. There are a number of useful anthropogenic tracers.

1. Trace Elements. It is now known that the proportions of various trace elements in pollution aerosols have distinct regional signatures that can be followed for thousands of kilometers, even though the concentrations of the individual elements decrease by an order of magnitude or more during transit. To date, time-series studies at high latitudes in the Northern Hemisphere have shown that the aerosols in remote regions reflect the incursion of air parcels with aerosols having a variety of identifiable, well-defined signatures rather than mixtures from widely separated source areas. On occasion, signals from strong point sources, usually smelters, are superimposed on these regional signals. Thus, trace elements are already useful as continental-and global-scale tracers.

2. Halocarbons. A wide variety of halocarbons that arise from industrial activities are released to the atmosphere. Several of these have been detected throughout the globe in the atmosphere, hydrosphere, and even the cryosphere. Some of these halocarbons have been identified as being uniquely associated with certain industrial areas. Many of these species have extremely long residence times, which makes them potentially useful as conservative tracers. Other species have well-known destruction mechanisms that make them useful for studying atmospheric chemical processes.

3. Elemental Carbon. Elemental carbon is produced in large quantities in combustion processes. Whereas industrialized nations are major sources, it is likely that even larger quantities are produced as a consequence of slash-and-burn agriculture. Elemental carbon can have serious environmental consequences because the particles are highly efficient absorbers of

radiation. Because the carbon particles are submicron, their atmospheric residence time is long. Elemental carbon is also relatively inert and, consequently, will not be chemically converted during transit.

4. Radioactive Isotopes (Man Made). Continental- and global-scale tracking of atmospheric motion became possible with the adve it of the nuclear age. Atmospheric nuclear testing injected large amounts of fission products into the upper troposphere and lower stratosphere. These products have been tracked not only through the atmosphere but also through a variety of other ecosystems affected by fallout. Although most atmospheric testing has been eliminated, not all nations are signatories of the Atmospheric Test Ban Treaty. We can expect to see occasional tests, and these opportunities should not be ignored.

C. Intentional Tracers

1. Perfluoronated Tracers. Perfluoronated tracers have been used in the CAPTEX experiments on a scale of 1000 km. They are very promising for experiments on a continental and global scale. They are described in detail in the presentation on gaseous tracers by Dietz and Senum (Section III).

2. Heavy Methanes. The utility of two isotopic analogues of normal methane, $^{12}CD_4$ and $^{13}CD_4$, has been demonstrated on continental scales of transport. An experiment is currently under way using these tracers to study circulation around Antarctica. Techniques are in hand for release and sampling of the tracer from surface sites and by aircraft.

3. Stable Isotopes. The heavy methanes are an example of stable isotopic tracers suitable for use on the scale of interest. Other compounds of unusual stable isotopic composition could, in principle, be used as well.

4. Radioactive Tracers. Radioactive species generated in atmospheric nuclear testing have long been used as global scale tracers. More recently, radioactive species released from nuclear facilities are being used on continental scales. Clearly, smaller scale releases of other radioactive species have potential on the scale of interest.

5. Physical Lagrangian Tracers. A physical Lagrangian tracer is an airborne instrumentation system that follows the flow of air in its vicinity (both horizontally and vertically), and can be tracked electronically. It is a constant-volume balloon system whose bouyancy is adjusted to follow mean vertical flows by using air as ballast. This is accomplished with appropriate sensors, an onboard microprocessor, pumps, and valves. It is an extension of constant-volume balloon technology that has been used extensively in both the stratosphere and the troposphere. In stratospheric applications, balloons have made multiple circuits of the earth. A light-weight, low-cost, satellite-tracked physical Lagrangian tracer is being developed for the Environmental Protection Agency as part of the National Acid Precipitation Assessment Program.

IV.   RECOMMENDATIONS

Systematically survey the tracer potential of elements and compounds presently being released anthropogenically.

Of the 60 or so elements that can be currently measured in atmospheric aerosols, fewer than 10 are being used actively as large-scale tracers. Of the tens of thousands of organic compounds being manufactured today, fewer than 1% have been sought in the atmosphere. Because anthropogenic tracers of opportunity are so promising, it would seem highly desirable to systematically survey the abundances, distribution, and tracer potential of elements and compounds presently being released into the atmosphere by man. Such a study would involve gases and aerosols, organic and inorganic species, and both sources and receptors.

Develop new tracer systems in which the tracer is subject to the same transformaton and removal processes as the species of interest.

Earlier, it was remarked that unreactive gaseous tracers with long lifetimes in the atmosphere simplify the study of transport and dispersion. The development of inert tracers is relatively well advanced. However, by themselves they tell us nothing about transformation and removal processes. To study these processes, it is necessary to have tracers available that are transformed and removed in the same way as the various species of interest--

be they sulfur dioxide, sulfate, acids, elemental carbon, smoke, or whatever. It should be a high priority to develop these reactive tracers because these processes are important to virtually every issue of interest.

Consider doing long-range tracer release experiments aloft in a Lagrangian frame of reference.

To date, most tracer experiments have involved releases at or near ground level and sampling on fixed Eulerian grids. On continental and global scales, however, the processes of principal interest are occurring aloft, as high as the mid to upper troposphere and even the stratosphere. To investigate these processes most efficiently, the tracers should be released where the action of interest is occurring. At these altitudes, Eulerian tracer experiments are difficult to perform. Lagrangian experiments, in which the measurement systems continuously follow the puff of released tracer material, are the approach of choice. Until recently, however, the technology required to do Lagrangian experiments was not in hand. The advent of both airborne LIDAR systems capable of remotely detecting fluorescent particle tracers and physical Lagrangian tracers with satellite tracking makes these experiments possible.

Develop radioactive species as tracers that are useful on an intercontinental and global scale.

Certain radioactive isotopes and compounds containing them are potentially the most sensitive tracers on this scale. This has been demonstrated by measurements of bomb test debris and recent measurements of ^{85}Kr released from Savannah River nuclear operations. Sufficiently large intentional releases of such radioactive species have not been performed to date and must be carried out with adequate regard to radiological safety; however, the amounts of material involved can be small (< 1 g), as can the amount of radioactivity relative to natural sources.

Typically, the release of 100 Ci of the tracer species should be measurable (> 10 x background) after dilution into one-thousandth of the global atmosphere by sampling 100 m^3 of air.

Among the potentially useful tracers:

$^{35}SF_6$	(half-life = 85 days)
$^{35}SO_2$	
$CH_2T\ CCl_3$	(half-life = 6 to 10 years)
$CTCl_3$	(half-life = 2 years)
$CTCl = CCl_2$	(half-life < 1 year)

The estimated half-lives are determined either by radioactive decay or by chemical reaction rates in the atmosphere. The development of each tracer system involves primarily the adaptation of established sampling and measurement techniques to these systems.

Establish a site for comprehensive sampling and research.

At least one site should be established where long-term measurements are made of a large number of atmospheric constituents. These measurements would provide a future "baseline" for the measured species. What is more important, they would provide a data base for discovering correlations and differential behavior and for formulating new hypotheses. This site should be a research facility as well as a sampling site. The participation should include as large a number of researchers as may be required by the diversity of constituents to be sampled and the wide range of analytical capabilities that will be required. Consideration should be given to locating this facility on an offshore platform on the continental shelf. Such a platform could also be used to study air/sea interaction, which has strong influence on air/sea exchange of aerosols and gases. At present, various sampling programs measure different species at different locations with different protocols and so do not provide much opportunity for discovering new relationships.

Formulate a national upper atmosphere and stratosphere program.

A national upper tropospheric and stratospheric sampling program snould be formulated that would incorporate ongoing independent research and sampling projects and programs. The program should include an improved ozone measurement program. Such a national program, under the guidance and patronage of a lead agency, would optimize the use of platforms, ensure proper intercomparisons and intercalibrations, increase the dissemination of data and results, increase communication between participants, and facilitate the interpretation of data through availability of more complementary measurements. Present research and monitoring projects sponsored by different agencies would benefit from multiple use of the same platforms. In addition, measurements taken under such auspices would compose a systematic national research and sampling program.

Encourage the development of an integrated global network of stations sampling for species that can serve as atmospheric tracers.

The spatial distribution of these stations, the species that are to be sampled, and the sampling frequency should be determined on the basis of the needs of the meteorological modeling community. Ancillary measurements should also be made of species that are important to atmospheric chemistry processes and to societal impacts. A prime consideration in the selection of species and sampling protocol is that they be based on relatively simple techniques that can be implemented at remote sites at relatively low cost.

Improve meteorology measurements over the oceans.

We recommend continued emphasis on the development of affordable in situ and remote sensing techniques for aerological observations. The acknowledged unsatisfactory distribution of routine upper air observing stations is a limiting factor in developing mechanistic and numerical models of atmospheric circulation. To adequately interpret and understand opportunistic and intentional experiments, better coverage over open oceans is needed. Appropriate technologies include soundings from ships in transit and aerosol sensing satellite systems.

Extend the scope of regional-scale intentional tracer experiments to the
continental scale.

   Realistic evaluations of the impact of anthropogenic emission upon acid
deposition, visibility degradation, and health, etc., requires the ability to
model the transport and dispersion of gases and small particulates over
spatial regions of 1000 to 2000 km in diversion. Calculations fail to give a
proper picture of transport on this scale because there is inadequate
knowledge of a variety of phenomena. Two examples of phenomena that
require further investigation are the effects of shear when material is
transported for time periods greater than 1-day and transport over complex
terrain.

   CAPTEX-83 demonstrated the capability of using perfluorinated tracers
up to a range of 1000 km. Future experiments should be done over greater
distances and cover a wider range of meteorological and terrain conditions.
Meteorological conditions should include stagnating highs, frontal storms,
etc. Terrain should include mountains (such as the Appalachins), the Great
Lakes, etc. The release methodology, sampling protocols, aircraft operation,
data processing, and program management of the CAPTEX-83 can be used as
a base upon which to build such experiments.

DRY DEPOSITION AND RESUSPENSION

I. CRITICAL RESEARCH QUESTIONS

A. Introduction

"Criticality" of research is defined by applications of the research results. The panel did not consider applications of dry deposition research explicitly, although it became apparent that many panel members were implicitly considering applications to the acidic-deposition issue. Nevertheless, the reader will probably conclude that a similar list of "critical" research needs would emerge from considerations of dry-deposition aspects of visibility deterioration, ecosystem stress by oxidants, fates of noxious materials released to the atmosphere, geochemical cycling of species, etc. In contrast, for the case of suspension/resuspension, the panel did enumerate a number of practical applications.

B. Dry Deposition.

This section is subdivided into three sections dealing with research questions about dry deposition of (1) particles, (2) gases, and (3) generic issues for both particles and gases.

1. Particles

a. Particle-size dependence of dry deposition to vegetative canopies. A succinct statement of the research question is: Does the well fill? For additional information, the reader is referred to Garland's Fig. 2 (Section III), which shows a minimum in the dry deposition velocity v_d for particles with diameters $0.1 < D < 1$ μm. These results, however, are for deposition to grass, both in the wind tunnel and (significantly) in the field. Yet the results are only for grass, and the question is: Is the minimum of v_d significantly less pronounced for deeper vegetative canopies? Theoretical results for deeper canopies suggest a negative response to this question (the predicted increase in v_d being, approximately, only in proportion to the increased leaf-area index). However, other field studies using both eddy-flux and

48

concentration-gradient methods suggest an affirmative response, especially for unstable atmospheric conditions. Neither response is overwhelmingly convincing because on the one hand, the theoretical studies are restricted to cases of horizontally homogeneous canopies, and on the other hand, the eddy-flux and concentration-gradient studies have failed to convince some researchers that the needed constant-flux condition has been satisfied. What is abundantly clear is the need for additional studies by a variety of experimental and theoretical methods, including additional evaluations of previous studies and new field measurements of tracer-particle deposition.

b. Sampling of supermicron particles of polydisperse particle-size distributions. The research question is: Are particle size-distributions (by number or by mass) being measured adequately for dry-flux estimates, especially the supermicron particles? There are a number of hints of inadequacies. If the number of supermicron particles has been underestimated in previous studies (for example, because of nonisokinetic sampling errors), then these errors might explain:

- the discrepancy between observed and predicted mass-average deposition velocities shown in Garland's Fig. 2,
- relatively large mass-average v_d values for SO_4^{2-} obtained by the concentration-gradient method, and
- the collection of more SO_4^{2-} on upward- than downward-facing collectors.

It is doubtful that particle tracers would be needed to resolve this issue.

c. Modifications of particle sizes, and therefore of dry deposition, caused by water-vapor condensational growth of particles. As yet, there have been no field studies of the effect of particle growth on particle deposition to lakes and oceans, and the use of tracers would seem appropriate. Some studies to examine the issue of particle growth have been performed in wind tunnels, but it is not clear that the most pertinent ranges of appropriate variables have been covered. Field studies with tracer particles would be valuable, especially in cases with fog and dew (super-saturated conditions); however, for such cases wind speeds are typically low, and the overall significance of particle growth to particle deposition (for example, for the acidic-deposition issue) may not be large, except for vegetation at high elevations. (We did not wish to become engaged in the question of whether this topic is dry deposition of wet particles or wet deposition of dry particles!)

    d. Supermicron particle bounce-off. Wind tunnel studies have demonstrated substantial bounce-off of supermicron particles from natural surfaces. The relevance of these results to field conditions, for 1- to 10-μm "pollution" particles and for larger particles such as those used in agricultural practices, needs investigation and presumably can be investigated relatively easily with available tracers. However, the microscopic properties of the tracer particles should be well specified.

    2. Gases.

    a. Are there additional suprises about the dry deposition of gases? The significance of this question can be seen from a few examples. When first uncovered, it was suprising that:

- NO_2 deposition to pine needles was several orders of magnitude larger than expected for a gas of limited solubility;
- Deductions of NO_x deposition, based on NO_x gradients, were complicated by fast NO_x-O_3 reactions in air and simultaneous O_3 gradients caused by its deposition;
- There is H_2 and HT consumption by biological activity in soils;
- There is a large O_3 dry deposition to vegetation and dry soils but not to water and wet soils.

Other topics where suprises might be found are: the deposition of NH_4NO_3 (in the presence of NH_3 and HNO_3) and the influence of isoprene and other reducing gases from vegetation on O_3 profiles and inferred deposition rates for O_3.

b. Surface-saturation/nonlinear effects on dry deposition. Specific questions include: (1) Why do SO_2 and O_3 depositions decrease substantially when stomates close, even though SO_2 and O_3 rapidly deposit on most surfaces? (2) How rapidly do surfaces saturate with SO_2, for example, during stagnant atmospheric conditions?

c. Influences of co-pollutants on dry deposition of specific species. Obvious examples include the influence of O_3 on NO_X deposition and the influences of H_2O_2 and O_3 on SO_2 deposition. For the latter case, it is noted that plant moisture typically has pH >7, and therefore O_3 is expected to be an important oxidizer of SO_2 within plants.

d. Dry deposition of hydrocarbons and free radicals. Research on these topics is warranted for a number of reasons.

- to determine the fates of released tracers (including the possible dry deposition of "nondepositing" species such as SF_6 and PFCs),
- to define the deposition of hydrocarbons whose ratios have been used to estimate atmospheric OH concentrations,
- for numerical models of hydrocarbon and free-radical concentrations.

e. Transfer velocities in surface media. For many gases (those whose equilibrium, air/surface-medium partition coefficient is smaller than that for DDT; for example, pyrene, SO_2 if pH of surface water <5, propyne, CO_2, etc.), their dry deposition (and resuspension) is strongly influenced by the transfer velocity in the surface media. Atmospheric scientists can contribute to the evaluation of these transfer velocities; for example, their dependence at sea as a function of wind speed (poorly known at present) and their dependence for soils as a function of humidity (and soil moisture). Tracers could be extremely valuable in these studies.

3. Particles and Gases: Generic Issues.

a. Surface inhomogeneities. A hierarchy of relevant questions is listed below.

- How accurate/inaccurate is the current practice of estimating total deposition to a mosaic of natural surfaces by simply summing the deposition to the component surface types?

- Can an upper bound on the error be found by determining the flux of momentum and/or moisture to inhomogeneous surfaces?

What progress towards the solution can be made by extending wind-tunnel studies of momentum fluxes to arrays of obstacles?

The use of tracer particles and/or gases to answer these questions appears to suffer from severe difficulties.

b. Dry deposition monitoring. The question of our ability to monitor dry deposition has been addressed in other workshops. A specific question posed here is: If there is concern that dry deposition to surface inhomogenieties may cause air concentrations to be horizontally inhomogeneous (generally expected to be a remote possibility for most pollutants), then can the concerns be alleviated relatively easily by exploratory measurements from a network of samplers in the neighborhood of a potential monitoring site?

c. Dry deposition in urban areas. Deposition of certain species (for example, I_2, HNO_3, SO_2, and a range of particulate materials) at specific sites in urban areas is of concern. Methods for determining even the concentration at sites within urban areas, with their many pollution sources, are not immediately obvious.

d. Adequacies of chamber and wind tunnel studies. This issue requires continued study by comparing results from these studies with those from field studies and through understanding of dominant processes.

C. Suspension/Resuspension

For reasons that are expected to become apparent, the group's considerations of suspension and resuspension differed from its deliberations on dry deposition. Thus, we first list a number of practical applications that are obvious.

1. Practical Applications.
- Resuspension of radionuclides from weapons testing, accidents, and previous routine releases
- Suspension/volatilization of noxious materials from waste disposal sites
- Vaporization of pesticides and herbicides
- Release of trace metals (organometals) from vegetation
- Suspension from the seas (of wastes and for estimates of geochemical cycles)
- Fugitive emissions from industrial activities and waste disposal sites
- Suspension of viable materials from municipal sewage treatment and other similar facilities
- Suspension and possible poleward-migration and distillation of hydrocarbons
- Influence of soil erosion on acidic/alkaline precipitation.
- Soil erosion

The fact that so large and important an application as soil erosion is but one of the practical applications of suspension/resuspension research may begin to convince the reader (as it did the group members) that this subject area is enormous.

2. General Comments. From considerations of practical applications of suspension/resuspension research, the following general comments emerged.

a. Inadequate funding. It seems clear, based on our knowledge of current suspension/resuspension research throughout the world (for example, as illuminated by the 1982 Chamberlain Conference), that there has been subcritical financial support of this research area during the past decade and that the support continues to diminish.

b. Lack of understanding. There is lack of understanding about fundamental chemical and physical processes involved in suspension/ resuspension (for example, "weathering rates," suspension from vegetation; partitioning of species among gas, liquid, and solid phases, etc.), and this lack of understanding severely limits extrapolations from "empirical data."

c. Interdisciplinary studies. Interdisciplinary studies are needed; many (if not most) should be led by representatives of disciplines other than meteorology.

d. Tracer needs. There is no clear indication that current tracer technology is limiting progress in suspension/resuspension studies. However, there are clear indications that more studies are needed to define the physical and chemical properties of tracers that are used.

e. Effect of atmospheric stability. Do stability and instability in the atmosphere give rise to greater changes in deposition rate than in momentum exchange?

f. Comparison of field methods. Several methods have been applied to measuring deposition of particles in field situations. Apparently there are large discrepancies in the results. A careful comparison of methods to evaluate the random and systematic differences between results given by the various methods would help achieve a better understanding of particle deposition in field conditions.

3. Some obvious research topics.
- Inhomogeneities of suspension caused, for example, by nonuniform terrain
- Fates of deposited submicron particles (for example, SO_4^{2-}: attachment to host particles; size distributions of resuspended material
- Correlations between soil erosion and acidity/alkalinity of downwind precipitation

- Suspension/resuspension from vegetation by winds (and associated mechanical abrasion), biological activities, and rain and environmental stresses
- Stabilization of spilled materials
- Suspension of noxious industrial materials from water bodies (influences of breaking waves, bubbles, etc.) and the susp⋅ :nsion of bacteria from sewage treatment facilities as a function of treatment methods
- Identification of the causes of modeling uncertainties (for example, "weathering rates"), initiation of research to reduce these uncertainties, and field studies to test the models

II. TRACER RESEARCH AND DEVELOPMENT OPPORTUNITIES

A. Resuspension

There appear to be two types of problems where tracers may be useful in understanding the physical processes. These problems include (1) locations where reasonably large amounts of toxic/hazardous materials have been deposited and may be entrained into the atmosphere and transported for some distance downwind, and (2) the general "erosion" of particulate material, which can have long-term effects both at the "erosion" site or downwind. The first category of problems comprises such man-made sources as coal piles, accidental spills of hazardous materials, and mine or mill tailings. The second category includes the resuspension of distributed materials such as fallout from atmospheric nuclear tests, the deposition of sulfate downwind from large SO_2 sources, and the erosion of agricultural soils. These are only a few of the possible applications where the suspension/resuspension process must be understood.

For many of these applications, a variety of tracers could be used. Some of the tracers are a portion of the deposited material, as in the case of a radioactive spill. In other instances, a tracer could be added to the source region of interest and used to monitor the resuspension process. The available tracers include (1) radioactive materials, (2) rare isotopes, and (3) a variety of chemical materials. The specific tracer requirements will depend on the application being considered.

Because of the broad range of potential problems and the variety of potential tracer materials available, there is need for better characterization of tracer materials and their behavior so that the chosen tracer is a more realistic surrogate of the original material. This characterization will require comparative tests of several tracers to determine their behavior and provide information that would be useful when selecting tracers for a specific application. Tracer characterization will provide generic background information that will improve our ability to apply existing tracers to problems of interest.

Another area of fruitful research is the long-term effect of resuspension. An ideal tracer is the radioactive material that has been released over the past 40 years. An integral part of this effort would be to map the activity to learn how the material is presently distributed.

B. Deposition

1. Particles. Particle deposition depends chiefly on three properties:

·   size, shape, and density as combined in the aerodynamic diameter,
·   hygroscopic properties, and
·   ability to adhere to the surface.

Investigation of the these can be pursued in both laboratory and field experiments by using tracers with a sufficiently wide range of these properties. Many such tracers have already been applied in the laboratory. There seems no need to seek additional tracers for laboratory use because those already reported in the literature provide good coverage.

Literature reports of field studies describe several tracers that are used in deposition measurements to show vegetation at modest range, but some of the above questions would be better investigated in experiments of greater scale and larger range. In particular, deposition to forest and urban deposition probably require a fetch of about 1 km, and effects of surface irregularities may require a fetch of ~10 km.

These greater fetches require that larger amounts of tracer be released, and some development of techniques is necessary to accomodate this requirement. The problem arises when particle number statistics and analytical sensitivities are considered. To avoid random errors in an experiment using monodisperse particles, a sample should contain 100 particles or more. When feasible air sampling rates and vegetation samples are considered, about 10^{11} particles must be released to allow measurements at a range of ~1 km, and about 10^{13} is necessary for a range of 10 km. Current methods for monodisperse particles allow ~10^8 particles to be generated in a reasonable time (for example, using a spinning top generator). Even if such numbers of monodisperse particles were available, their small mass would give rise to analytical difficulty when small particles are studied. The mass of one hundred 1-μm particles is about 0.1 ng and the mass of 0.1-μm particles is only 0.1 pg.

As an alternative to monodisperse particle experiments, it is possible to generate a large quantity of tracer in a polydisperse aerosol, and then examine both air and deposition samples by a technique capable of determining the numbers of particles in limited size ranges. Measurement by an SEM equipped with x-ray analysis and automatic image analysis is a possible method. Grams or kilograms of tracer can be released using atomizers, pyro-technique, and other methods that are sufficient for experiments at ranges of many kilometers.

Dual tracer techniques are attractive possibilities for determining total deposition over such scales by using plume depletion modeling. However, the method is not applicable to heterogeneous terrain unless sufficient measurements can be made to estimate the mass balance of the plume--and at present this is barely feasible.

For large particles, particle mass is large enough that adding a small fraction (by mass) of tracer is sufficient. Spores and pollens, much used in the past, and modern, uniformly sized microporous materials manufactured for use as supports in chromatographs are among the possible particles. Some high molecular-weight freons are involatile and can be analysed with high specificity. They join the dyes and radioactive and trace-element tracers that may be used to identify the particles. In the >5-μm-diam range, the question of bouncing is important. Comparison of the deposition of liquid droplets (nonvolatile oils labeled by the same range of tracers in solution or suspension within the droplet) with that of the solid particles already mentioned is a recognized technique for investigating the bouncing of solid particles. Production of sufficient numbers of such droplets for field experiments probably requires further development.

2. Gases. The measuring deposition of reactive gases to natural and surrogate surfaces is an important research problem. At this time there are few definitive ways to measure these species except in laboratory studies. A number of specific gases are of interest, including the oxides of sulfur and nitrogen, ozone, halogens, and hydrocarbons. It is recognized that stable and radioactive isotopes of gaseous elements offer the possibility of definitively tracing the rate of reactive gases in natural environments. However, it may not be possible to sample a sufficient mass of the species from natural surfaces to perform the measurements using stable isotopes. This problem could be addressed by the use of radioactive ^{35}S, but the perceived health risks will probably preclude its use. Applying stable isotopes of sulphur is possible if sufficient quantities can be produced and released economically.

One area that is potentially very important for understanding gaseous deposition is the effect of vegetation surfaces. For example: How do leaf surfaces affect the deposition rate and uptake of reactive material? This problem can probably be addressed by the use of radioactive ^{35}S because of the high content of sulfur in living matter.

Using surrogate tracers for reactive gases does not appear to be a viable option unless the chemical reactivity of both gases is well understood.

III. TRACER NEEDS

For particle deposition, we need one or more monodispersed tracers that can be released in large quantities, for example, 10^{11} particles over periods of 5 minutes to several hours. This may require a new mode of tracer release instead of, or in addition to, actual new tracers. Sizes ranging from 0.1 to 3 μm or more are needed.

We need a complimentary approach in the development or improvement of techniques to count particles as a function of particle size. This would allow the use of polydisperse tracers that can be released in substantial quantities.

It would be useful to have both depositing and nondepositing tracers on which one could perform mass budget measurements by using remote sensing techniques. The development of such techniques is also necessary.

Although there are still numerous problems in understanding and describing the suspension of material from various surfaces, the availability of suitable tracers does not appear to be a major stumbling block at this time.

REGIONAL AIR QUALITY

The <u>regional</u> scale of transport and dispersion problems can be defined as extending from the <u>local</u> scale (0 to 10 or 20 km) on the lower end, to the beginning of the <u>continental</u> scale (perhaps 1000 km) on the upper end. The idea of a regional scale relates primarily to air pollution regulation (air quality regions), but it includes the entire <u>mesoscale</u> (tens to hundreds of kilometers) of atmospheric phenomena. The regional scale is the most incompletely understood of all the scales of atmospheric motions, including a variety of meteorological phenomena that strongly affect air quality.

For the purpose of identifying and studying tracer applications to regional air quality problems, the various regional scale atmospheric phenomena have been grouped as follows:

- complex terrain effects,
- urban-rural transport and dispersion,
- coastal effects,
- cloud processes and organized convection system effects, and
- dispersion of hazardous materials.

These problem areas will be discussed in relation to

- local and regional scales of transport and diffusion and their interactions,
- general meteorological and detailed chemistry and physics effects,
- tracer applications to (1) problems of regional air quality and (2) understanding the atmospheric processes involved in regional-scale transport and diffusion.

Within this framework, the intention is to identify and characterize useful tracers--those in common use and those whose development is desirable and feasible--and, for the latter group, to consider their release, sampling, analytical, and experimental design characteristics.

60

I. COMPLEX TERRAIN

In many cases, large energy resources are found in areas of complex terrain that is of nonflat orographic configuration, including the effects of mountains, lakes, and rivers. The development of these energy sources results in emissions of atmospheric pollutants. To develop these sources in an environmentally acceptable manner, the air quality impact of source emissions on scales from local (~10 km) to regional (hundreds of kilometers) must be understood. Previously, atmospheric boundary layer research was largely confined to relatively flat terrain. Only recently has the atmospheric science community begun to investigate atmospheric boundary layer flows over complex terrain in an intensive and organized fashion.

A. Past Use of Tracers

Previous atmospheric boundary layer studies in complex terrain that involved tracers have been generally confined to the local scales (10 to 20 km). At first these studies were confined to source receptor relationships, and little attention was paid to the dynamical processes occurring between the sources and receptors. From these initial experiments, it became clear that the transport and dispersion of material in complex terrain was not easily explained by models available at that time. In the past several years, more sophisticated experiments have been designed and executed that use tracers to help investigate the fundamental dynamical process involved in complex terrain boundary layer flows.

Nocturnal drainage winds have been studied using tracers in conjunction with meteorological sensor arrays to investigate the development, merging, pooling, and breakup of these flows. Also the coupling between the drainage wind and the flow above has been investigated using tracers. In these studies, multiple tracers were simultaneously released and monitored. In addition to clarifying the dynamical processes, these studies provided data on the integrated transport and diffusion processes in complex terrain.

Other complex terrain studies have used tracers to investigate atmospheric flows around isolated hills, over ridges, and within isolated valleys. These studies generally have been confined to the local scale; however, the scale of interest is expanding as understanding of local-scale processes improves.

B. Future Uses

Moving from the local to more regional scale in complex terrain, the focus is on (1) developing an understanding of interaction between local circulations on different scales occurring within individual valleys, (2) the interaction between adjacent valleys, (3) merging of flows from several valleys, (4) coupling between valley flows and the regional flows, and (5) transport and diffusion over distances of several hundred kilometers with a complex terrain lower boundary condition. As both the horizontal and vertical scales of these studies expand, the demand placed on present tracer release, measurement, and analysis techniques increases and new requirements are identified.

C. New Requirements

Given the complex terrain meteorology research needs discussed above, tracers must be a versatile tool that can be used in conjunction with basic meteorological investigations and models to gain a better understanding of the basic physics of complex terrain meteorology on both the local and regional scales.

The most immediate benefits to complex terrain investigations are expected to come from the use of multiple, nonreactive, gaseous tracers. Much of the tracer technology necessary in local-scale complex terrain investigations has already been developed. Improvements and modifications to existing tracer release systems, tracer sampling and analysis systems, and equipment will be the primary needs for local-scale investigations. In contrast, significant research and development effort will be necessary to obtain a useful tracer technology for regional-scale meteorological investigations in complex terrain. Regional-scale investigations are expected to require a larger number of tracer materials than have been used to date and will also require advances in the development of a real-time, continuous tracer analyzer that can be carried by aircraft. In addition, further development is needed to produce a tracer system that can be used to obtain information on long-time average concentrations (for example, annual average concentrations), as required in certain regulatory applications where detailed understanding of physical processes is relatively unimportant. A promising perfluorocarbon technique has been suggested for this use. In the remainder of this section, we will discuss tracer technology needs for shorter term average concentrations, as needed in research designed to provide a better understanding of the basic physics of complex terrain meteorological processes.

D. Release Technology

Workshop participants identified several improvements and innovations in tracer release systems that would be desirable in complex terrain investigations. (1) A better means of conducting elevated releases of tracers should be developed for the case when a tower is unavailable or is prohibitively expensive. At present, there is no means to release a particulate tracer for visualization of the gaseous plume in such circumstances. Establishing the release point is often difficult when tethered balloons are used for the release. (2) New dual-tracer techniques should be developed to allow an investigator to determine the release time of a collected sample. This would be useful in complex terrain investigations where air parcels can follow complicated trajectories before they are sampled. (3) Procedures and techniques should be developed to release tracer material into complex terrain circulations of different scales (for example, slope flows or along-valley flows) to study scale interactions. This will require release systems that are more portable and flexible than any available to date.

E. Sampling Technology

Several specific needs were identified by workshop participants. (1) Vertical sampling devices are necessary to provide 3-dimensional information on tracer plumes in complex terrain areas. Further research and development work is necessary to satisfy this need, although some has already been done (most notably at Sandia National Laboratories) on samplers that can be carried by tethered balloons. (2) Modifications to existing sampling devices and development of new sampling devices are necessary to gain flexibility in tracer sampling in complex terrain. To be used on valley sidewalls or in rather inaccessible remote areas, the sampling devices should be portable, battery operated, radio controlled, and capable of sampling at different flow rates and over different averaging intervals. (3) A means of remotely sensing gaseous tracer concentrations in a complex terrain area should be developed. This technique might be similar to the COSPEC technique already developed for SO_2 and NO_2. (4) Continuous real-time samplers should be developed for use with the new perfluorocarbon tracers so that regional-scale transport and diffusions investigations using aircraft can proceed.

F. Analysis Technology

Several suggestions were made by conference participants concerning tracer analysis procedures.

(1) Techniques, equipment, and procedures should be developed so that concurrent analysis of multiple tracers can be accomplished from single air samples by using a single-analysis device.

(2) Quality assurance procedures should be improved to provide additional information on the quality of samples. Redundant data should be collected in the field (for example, two samples could be taken concurrently at the same location, or two samples could be taken in the same vicinity) to determine if analysis errors are occurring or if a single site is representative of other sites in the vicinity.

(3) For tracers whose analysis is done in the laboratory long after the field experiments are completed, procedures and real-time analysis equipment should be developed to verify that concentrations are in the range expected and that analysis errors are not occurring as a result of processes taking place between sampling and analysis.

G. Recommendations for Future Needs

The major needs for improved tracer technology identified by the panel are

- •   better vertical release and sampling mechanisms;
- •   real-time analysis of tracer samples; and
- •   multiple (3 to 4) tracers that can be released simultaneously, collected by one sampler, and analyzed in one process.

Releases of tracers in the vertical direction should be available to a minimum of 1 km and have the flexibility of releasing more than one tracer from a single or multiple points in puffs, plumes, or line sources. The release heights should be specified and implemented near the time of the experiment. Vertical sampling should be available to a minimum of 1 km and have the flexibility of defining approximately 10 measurement heights with a minimum resolution of a few meters. The capacity for sequential measurements at each level is also a requirement.

To reduce the logistics for multiple-tracer releases, we recommend that multiple tracers (3 to 4) be available for simultaneous collection in one sampler and analysis by one procedure. This will reduce the cost of multiple-tracer releases and the chance of inconsistencies and errors.

Real-time analysis of tracers is required (1) to determine whether a release is behaving as expected so adjustments to the experimental plans can be made for subsequent experiments, and (2) to track and measure regional-scale tracer releases by aircraft or ground vehicles.

In general, we recommend that careful design of experiments be made a top priority. This implies developing a plan that includes initial specific goals, design and execution of an experiment, and specifics of analyzing and interpreting the data.

Another area that needs further study, particularly in complex terrain, is the use of tracers to help define the spacial and temporal representiveness of measurements. This involves co-location or near co-location of samplers and the appropriate meterological instrumentation to help define observed variabilities in concentration measurements that are caused by natural variations in the atmosphere and possible variations in the samplers. In addition, the representitiveness of the release site should also be considered when designing an experiment.

II. URBAN/RURAL SCALE OF TRANSPORT

Concern for air quality covers regions located within urban boundaries and in surrounding rural areas. These regions are influenced by both local and distant sources. By definition, then, the scale of transport being considered is from local (a few kilometers) to long range (up to 1000 km).

On these scales of transport and dispersion, meteorological parameters have significant influence on pollutant concentrations. During their passage through clean air, rainfall, and cloud systems, physical and chemical effects will cause further change in the type and concentration of pollutants remaining in the air and on their removal and chemical conversion rates.

Our understanding of the details of these processes that occurr on regional scales is poor in the following areas:

- urban heat island effects on pollutant transport and dispersion from both within and outside this zone of influence,
- the impact of long-range transport and dispersion from point and regional sources on receptor areas as much as 1000 km away, and
- source attribution--what portion of a pollutant measured at a receptor site is attributable to a distant pollutant source.

Other influences on these scales, such as local and mesoscale complex terrain, coastal meteorology, cloud processing, organized convective systems, and atmospheric dynamics are important and are being covered in other sections.

Methods are needed for determining the origins, both by class of source (for example, coal-fired plants, motor-vehicle emissions) and by area (for example, Ohio River Valley, Chicago Metropolitan area, Sudbury region of Ontario), of various chemical species in the particulate and gas phases that are observed at a particular location and time. With this information about sources, it would be desirable to know how the original distribution of species at the point of origin was modified during transit by dispersion (both horizontal and vertical), by deposition, and by transformation, both chemical and physical.

To understand the influential mechanisms in these problem areas, we must make use of intentionally released conservative and nonconservative tracers, both gaseous and particulate, as well as different trace constituents inherently present in various pollution sources (that is, the so-called tracers of opportunity). Experiments should be considered in which these tracer tools are used separately, and in coordinated simultaneous release studies because the dual approach offers the potential for better interpretation as well as reduced costs in implementation.

Because inherent tracers and intentionally released tracers are different systems comprising type, release, sampling, and analysis components, they will be discussed separately.

A. Priority Needs

1. Conservative Gaseous Tracers (Intentionally Released). Considerable laboratory research and field implementation has gone into this class of tracer. The research needs will be discussed briefly in terms of type, release equipment, sampling, and analysis.

a. Type and Number of Tracers. Although numerous types (for example, SF_6, halocarbons, perfluorocarbons, heavy methanes) are available, on the scales of transport considered here, the heavy methanes do not appear to be viable. The SF_6 and halocarbons are extremely useful and effective on scales up to 10 to 50 km or so when, for example, single, or at most dual, tracer experiments involving sampling on the ground and aloft are desired. Real-time information is available with SF_6 continuous analyzers.

Perfluorocarbon tracers (PFTs) are the most useful when more than one type of tracer is desired, because the sampling and analysis can all be conducted with the same equipment. For long-range applications, only the PFTs are cost-effective when many sample analyses are required. However, at present, there are only three PFTs available.

Specific needs include

- establishment of at least five or six PFTs and
- a center for tracer coordination.

Research must be conducted for new PFTs including tests for purity, suitability in instrumentation, field use, and cost. A center should be established, supported by government and the private sector, to oversee the development, production, and coordinated use of PFTs (and perhaps all tracers). This pool of resources would provide greater economic power for tracer production and quantity purchases.

b. Tracer Release. Ground and vertical releases of tracers are needed in continuous, pulsed, and possibly puff modes. These may be required as point, line, and regional sources. Equipment must be reliable and automated if, as in some cases, it may have to operate for up to 2 years in a preprogrammed fashion.

Continuous releases emulate steady-state emission sources; pulsed releases may provide a means to determine downwind travel time and along-wind dispersion; and puff releases simulate accidental spills or explosive sources. An important consideration is the method needed to tag a moderately sized (up to 100-km^2) region with a tracer to simulate ground and aerial sources.

Specific needs include

- release equipment automation, flexibility, and reliability;
- modes of release: continuous, pulsed, and puff;
- vertical releases: stacks, balloon-borne cables, aircraft; and
- studies of line- and area-source tagging.

c. Tracer Sampling. Sampling components to be considered include location, frequency, duration, and quantity as well as types of samplers. Specific improvements needed include:

- reliability and automation of the programmable PFT samples,
- increased quantity of samplers for greater resolution in temporal and spacial field sampling,
- greater speed and resolution in real-time PFT sampling, and
- validation of vertical sampling techniques.

Real-time remote detection (for example, lidar) is not now attainable nor will be in the next few years.

d. Tracer Analyses. The desire to use an increased number of PFTs and the quantity of samples collected in a given experiment place burdens on the analysis systems. Specific research needs include:

- studies on improved resolution gas chromatographic (GC) techniques,
- automation of analyses and data handling, and
- incorporation of tracer ratioing techniques for greater precision and accuracy.

2. Nonconservative Tracers (Intentionally Released). The tracers in this group include gaseous water-soluble tracers, gaseous reactive tracers, and scavengeable particle tracers. Many of these interactive tracers are in the conceptual stage, and others are perhaps at the engineering and/or proof-of-concept stage.

a. Water-Soluble Tracers. To emulate the fate of water scavengeable pollutants in interactions with clouds and rain, it is desirable to have a tracer that is sensitively detectable but that also has a known solubility in water. Atmospheric freons are a class of such compounds that have good sensitivity to detection by an electron-capture detector (ECD) but generally have a low reversible solubility. Another possibility are freons that form hydrates such as freon 12B1; unfortunately, most also seem to have a low, reversible solubility. Certain fluorinated/chlorinated ketones hydrolize to form genidiols; hexafluroacetone has a hydrolized/unhydrolized ratio of 10^6 and an ECD sensitivity about 1/60 that of SF_6. Thus, it and similar ketones should prove to be good soluble tracers. Aldehydes are expected to behave similarly. Finally, perfluorocycloalkyl acids might be the best choice because, on hydrolysis, they would ionize in solution. Thus, the acids would form a class of irreversible water-soluble tracer.

Specific research needs include:

- detailed study on solubility of perfluorinated ketones, aldehydes, and acids;
- development of appropriate analytical schemes; and
- determination of the rates of hydration and any potential toxicity.

b. **Gaseous Reactive Tracers.** Pollutants, organic and inorganic, undergo oxidation processes in both the gas (for example, OH and NO_3 radicals and O_3) and aqueous phase (for example, through H_2O_2, etc.). Specific needs include:

- a review of ECD-sensitive compounds that undergo selective gas- and aqueous-phase oxidation, and
- a review of compounds that undergo such processes and whose products can be readily converted to ECD-sensitive compounds.

c. **Scavengeable Particle Tracers.** A significant amount of research and field implementation has gone into the use of intentionally released particulate tracers, including oil fog, fluorescent particles, pollens and spores, dyes, and rare-earth compounds, each of which have their own specific sampling and analysis schemes. Recently, several rare earths enriched in normally minor isotopes (for example, neodymium) have been proposed as long-range tracers of particles emitted from smoke stacks; they would be detected by sensitive isotopic ratio mass spectrometry. Specific research should be directed toward techniques to uniquely identify particle tracers for use in mesoscale (10- to 100- km) and long-range (100- to 1500- km) studies.

3. Inherent Tracers. Considerable progress is being made in the use of inherent tracers, "tracers of opportunity," to identify the contributions of particles and gases from certain types of sources; these contributions are made to the ambient air in urban and rural areas and/or source areas on a regional basis. Tracers of opportunity are materials emitted during normal operation rather than intentionally added tracers. Most work in this area has involved measurements of many elements borne by particles, but other kinds of species are being developed as tracers: organic compounds, vapor-phase species (for example, organic compounds and compounds of B, Se, halogens, Hg), morphology and composition of individual particles, and isotopes of certain elements, both stable and radioactive (for example, ^{13}C, ^{14}C, ^{10}B, ^{11}B, ^{34}S, ^{35}S, ^{36}S, natural heavy-element activities). The use of particle-borne elements on an urban scale is fairly well developed and is already (for example, in Portland, Oregon) for making regulatory decisions. Considerable research is still needed to develop and test the use of particle-borne elements for regional-scale applications and other species on all distance scales. Specific items of interest, which are discussed in detail below, include:

- more definitive source signatures,
- detailed characterization of different sources under many conditions,
- detailed particle size and composition,
- vertical profiles of particles and gases, and
- coordinated inherent/intentional field tracer studies.

a. Definitive Source Signatures. The elemental compositions of particles from a few kinds of sources, especially coal-fired power plants, have been rather thoroughly studied, but measurements are needed on a much wider range of sources and for many additional types of species. Sampling should be done in such a way that, as nearly as possible, the results represent the chemical and physical states of the released materials <u>as they would be perceived at distant receptor sites</u> (for example, instead of sampling hot stack gases, use dilution source sampling or sampling at some distance downwind in the plumes). New information in this area will be available soon as a result of the EPA-sponsored studies in Philadelphia during July/August 1982, the EPA/DOE/EPRI-sponsored studies at Deep Creek Lake, Maryland in the summer of 1983, and the correlated studies by Ford Motor Company at Allegheny Mt. and Laurel, Pennsylvania.

b. Concentrations of Many Species at Rural Sites Under Many Conditions. Very few thorough measurements of gas and particle compositions have been made at rural sites and then correlated with meteorological data (for example, back trajectories) to determine characteristic patterns associated with certain regions or sources. Much more data of this type is needed to determine whether or not such signatures can be established.

c. Detailed Particle Size and Composition Data. The results presented by Gordon (Sec. III) indicate considerable fraction of particles, even those with diameters 2.5 μm, during regional-scale transport. Studies by Ondov have revealed considerable fine structure among particles of 2.5-μm diameter in the emissions from coal-fired plants and in ambient air. Further studies should be done to exploit the information content of these very fine particles.

· Vertical Profiles of Particles and Gases - Nearly all of our present information on detected compositions of particles and gases is confined to ground level. To gain information on long-range transport and dispersion of these species, we must have much greater knowledge of vertical concentrations and movements of various species under a wide range of meteorological conditions. Such studies would be especially useful if conducted downwind of dominant sources (both elevated and ground level) of certain types of species, for example, coal-fired plants or smelters, regions of high traffic density. If we had better knowledge of the concentrations of many types of species at cloud-forming levels, we could use them as tracers of processes that occur in clouds, for example.

Correlation of Intentional Tracers with Tracers of Opportunity To test and confirm the use of tracers of opportunity, links
should be established between the two types of tracers.
Intentional tracer releases should be done near similar point, line,
or area sources that have good tracers of opportunity associated
with them. Collection of the intentional tracer should be
accompanied by collections of many other species that may be
good tracers of opportunity.

B. Recommended Future Research

Tracer studies of regional transport and dispersion over distances up to
1000 km are needed to resolve questions relating to long-range transport of
pollutants, including those contributing to acid deposition. Ideally,
commitment to a 10-year research plan involving a series of CAPTEX-like
requirements (perhaps every 2 years) is recommended. The atmospheric
science community (government, private industry, and universities) should be
invited to participate. Modelers and research scientists should participate in
an organized way to make use of the experimental data for model testing and
evaluation. Most important, after each series of field experiments there
should be feedback from the research community to improve the design of
follow-on experiments.

Experimental data are needed for a variety of meteorological
conditions; for example:

- tracer releases in the vicinity of frontal zones with strong
  vertical motions and precipitation,
- releases into a stagnating high-pressure system,
- night-time releases of different tracers at ground and stack
  height to observe the influence of stack height on long-range
  dispersion, and
- in addition to diagnostic studies individual plume dispersion, there
  should be long-term climatological studies of transport and
  dispersion.

Tracers could be released on a regular schedule over a period of months or perhaps a full year. Monthly average tracer concentrations could be measured at a limited number of sites at a range of distances from the release points to determine the integrated long-term average effects of transport and dispersion; these data then could be compared with dispersion model predictions. Both long-range experiments and local scale experiments, such as the METREX (Metropolitan Tracer Experiment) now in progress in the Washington D.C. area, are recommended.

The panel recommended future work be done in the research areas listed below.

- More definitive source signatures should be developed for a wide range of types of sources.

- Much more rural sampling should be done and correlated with meteorological data to develop and test source type and area attribution schemes.

- Much more attention should be paid to the composition of particle-size groups of 2.5-μm diameter . Some research is probably needed to develop more reliable particle-size separation devices for this size and those with high flow rates.

- Vertical distributions for many gasand particulate-phase species are badly needed to gain better knowledge of three-dimensional transport and dispersion. Whenever sampling flights occur, provision should be made for thorough sample collections for this particular purpose unless they are incompatible with the major aim of the flights.

- Whenever intentional tracers are released, the experiments should be planned in such a way that they simultaneously test the use of tracers of opportunity; for example, the releases should be made close to dominant sources of tracers of opportunity and the latter species should be collected with the intentional tracer as often as is practical.

It is, or course, implicit in each of these recommended research areas that the proper tracer tools, as calculated earlier, be developed and field tested. This includes development of:

- new conservative PFTs,
- an improved programmable sampler,
- high-resolution and high-speed analysis equipment, and
- high-volume particle collectors that subdivide in <2.5-μm diameter particles.

III. COASTAL TRANSPORT AND DIFFUSION

Coastal regions are characterized by significant problems involving the transport and diffusion of atmospheric pollution. The concern originates with the complexity in atmospheric dynamics that results from abrupt topographic changes that occur at the land/water interface. These changes lead to unique local-scale wind circulations such as the seabreeze and its diurnal counterpart, the land breeze. The thermal internal boundary layer (TIBL) likewise is a local-scale phenomenon that has profound impact on the transport, diffusion, and distribution of pollutants, either those emitted in the coastal zone or those advected into it. Other local-scale phenomena may also have unique effects on the local to regional-scale transport and diffusion in coastal zones.

Numerous studies have been conducted in recent years to describe, define, and predict the atmospheric behavior, including atmospheric pollutant behavior, in coastal zones. The atmospheric tracer is a useful tool in these studies, but in the past, limited experimental methods have resulted in certain observational limitations, which have in turn limited development of our understanding of the atmospheric dynamics that control pollutant behavior. To more fully develop our understanding of the coastal physical atmospheric processes and the means to predict pollutant behavior, it will be necessary to improve experimental techniques. This discussion attempts to identify the primary phenomena that must be further investigated, to suggest tracer methods that could be employed in such investigations, and finally, to identify current areas where those tracer methods should be improved to provide the most useful experimental methods.

A.   Coastal Transport/Diffusion Mechanisms

The primary mechanisms that control pollutant transport and diffusion in coastal regions are the sea-breeze/land-breeze regimes, diurnal cyclical flows (including recirculations), the thermal internal boundary layer (TIBL), and low-turbulence ducting (over water).

The seabreeze/landbreeze and various forms of local-scale wind flow phenomena such as helical flow patterns and the flows that have diurnal, cyclical aspects result in recirculation of pollutants. Sometimes this involves recirculation back to the region of origin and thus results in unusually high concentrations. At other times, the results may be advection of pollution to areas that would otherwise not expect to be exposed to the pollutant at all. The presence of the TIBL results in unusual behavior of plumes, so rapid "fumigation" can occur over regions of limited size; this may result in very high concentrations of pollutants for short periods of time. Zones of low turbulence, generally located over water, can result in plume-trapping. While the plume is within this zone, it experiences very low diffusion rates and thus retains comparatively high concentrations.

B.   Experimental Methods Needed

One of the primary reasons that little is known about atmospheric dynamical processes in coastal regions is the limitation of experimental methods. Among the most prominent of these limitations is the fact that readily useable, stable, over-water instrument platforms have not been available. The result is a lack of detailed measurements over water. Data sets that might characterize unique coastal atmospheric phenomena simply do not exist or are not as complete as those for over land. To more fully understand transport and diffusion of pollutants in coastal zones, we need certain advances in atmospheric tracer systems. Some key areas requiring development are listed below.

- A capability is needed for tracer sampling at surface level over water.

- A capability is needed that would yield vertical distributions of tracers. At present, this can be partially satisfied by real-time aircraft sampling, but more detail is required than can be obtained now by aircraft sampling approaches. Vertical profiles are needed at multiple locations, at a sequence of times, and for long periods of time--most probably to a degree that would be very expensive if attempted by aircraft-mounted sensors.

- Perhaps the two above requirements point to the need to develop remote sensing capabilities for tracer elements. Some pioneering steps have been made in this area and might usefully be pursued for further development (for example, observation of paint pigment with partical LIDAR).

- Multiple tracers are needed that would have essentially the same sampling and assay approaches and the same source-to-sample sensitivity and which could be assayed in a relatively short time to provide thousands of samples per day.

- Long-term sampling capabilities are needed to explore the dynamical aspects of unique coastal phenomena such as the TIBL. In this case, emissions could be made at different positions and times with respect to the TIBL, and subsequent ground-level samples and vertical profiles could be employed to describe the physical processes.

- Tracers of opportunity may be particularly useful for studying processes in the coastal zone because many activities that generate such "tracers" are located in coastal regions. Proportionally, the greatest number of people live in coastal zones, and a large portion of the manufacturing, power generation, and other industrial activities are carried out there. These activities generate material that might be employed as a tracer of opportunity. Such techniques should be developed as a means of supplementing other tracer techniques.

IV. CLOUD PROCESSES

Clouds can both transport and transform pollutants. Strong convective systems carry pollutants from the surface mixed layer to the free troposphere where high wind speeds facilitate long-range transport. Tracers can be used to qualify the process and possibly quantity it.

Gases and particles contained within the mixed layer may undergo processing by short-lived cloud many times. Specific aspects of the cloud/pollutant interactions require study to improve basic scientific understanding. Because regional scales may be quite different in terms of both pollutants and cloud types, one must examine the result of these interactions as they occur over the scale of hours or several kilometers. Tracers, in these applications, serve as identifiers of air parcels so that the chemical and physical evolution of the pollutants may be examined.

The subsequent discussion of chemical and physical processes in the cloud environment considers some of the aspects that might retard regional air quality but avoids wet deposition, which was discussed by another panel. Precipitation systems are not covered, but they may improve regional air quality and have important air motions that are worthy of study.

A.   Chemical and Microphysical Processes

The following cloud chemical and microphysical problem areas that may impact upon regional air quality have been identified:

(1)   Nucleation - What are the nuclei for cloud droplets, how do they influence aqueous-phase chemical processes, and does dissolution and evaporation affect the nuclei?

(2)   Gas to Particle Conversion-To what degree does absorption of
reactive gases by the cloud droplets and later evaporation of
these droplets affect the aerosol particles?

(3)   Evolution of the Submicron Particle Distribution- Do unactivated
aerosol particles coagulate with cloud droplets, and are clouds
significant to the transferal of Aitken nuclei to large aerosol
particles?

(4)   Entrainment - Does the entraining of air containing reactive gases
significantly enhance the conversion of these gases to particles,
and do the entrained nuclei participate in cloud droplet
formation--thereby providing additional centers for chemical
reaction?

(5)   Degree of Material Processing-How much of the air in the mixed
layer is processed by cloud, and for how long? In terms of a
regional-scale problem, this might be based upon the time taken
for air to be horizontally transported over this scale (that is,
hours).

B.   Tracers for Chemical and Microphysical Processes

Identification of the cloud droplet nuclei cannot satisfactorily be
handled by tracer technique because it would require tagging the cloud
droplets in-cloud without contamination of the interstitial particles.

The consequences of dissolution and evaporation, gas-to-particle
conversion, and the evolution of the particle-size distribution can be tested in
several ways. A nonreactive gaseous tracer may be used to mark an air
parcel that is expected to participate soon in cloud formation. Appropriate
measurements of gases and particles in the pre-cloud "tagged" air and in the
"tagged" air after cloud evaporation could be used to evaluate problems
(1)-(3). Such an experiment requires a detection system with a fast-response
(that is, seconds), real-time measuring capability. Consideration must be
given to the effect of dilution of the air parcel through mixing processes both

in the cloud (for example, entrainment) and after cloud evaporation. Concentration measurements of the nonreactive tracer might be used to quantify this effect; however, the experimental design must be well planned.

Particulate tracers are another means of addressing these problems. An experimental design similar to that above might be conceived, but a tracer that can be detected in real time is again required. A second tracer release for purely "marking" purposes could be used. A particulate tracer has the advantage that the processes of interest may be examined with considerable accuracy if the tracer is distributed in a particle size that mimics the ambient size distribution. Those considered for particulate tracers are enriched rare-earth isotopes, isotopes of sulphur, and the inherent "tracers of opportunity." The limitations are the requirement of a large sample volume, adequate representation, the pollutants of concern, and dilution of cloud "processed" air with "nonprocessed" air.

To answer the questions concerning entrained air stated in (4), reactive tracers (both gaseous and particulate) might be released in regions surrounding the cloud and then sought in the cloud water. This study would be assisted by the simultaneous release of a nonreactive tracer that is detectable in real time and has a fast time response. It would not be necessary for the reactive tracers to be detectable in real time; a puff or line release may be most appropriate. It is not likely that quantitative information can be gained from the reactive tracer; however, the identification of the entrained air's origin is useful to quantitative assessment by other techniques. One such technique is through the conservation of particle-number concentration in the accumulation mode. These concentrations are conservative over short periods, and in the absence of significant sources (that is, plumes), the concentration of cloud droplets plus aerosol particles in a cloud element may be related to the cloud-base concentration of aerosol particles and the concentration in the entrained air.

The question of the degree of material processing may be answered by comparing the individual cloud results with those obtained from longer time scales, but it is probably better addressed by the simultaneous use of a nonreactive and reactive particulate and/or gaseous tracer. The reactive tracer should be either destroyed or somehow marked by the cloud environment. At least one tracer (the nonreactive tracer) should be identifiable in real time.

The study of these problems using tracers has been, out of necessity, oversimplified by the preceeding discussion. In actuality, the design and logistics of this type of experiment, particularly with aircraft, require careful and extensive planning. This factor should not be overlooked when considering the application of tracers.

To briefly summarize the tracer requirements of this section, there is a need for at least one nonreactive tracer system than can detect in real time with response times of seconds. Using more than one tracer can provide additional information and improve resolution of the problems. Reactive tracer applications should be considered in more details in order to evaluate what information can be extracted. Finally, tracer concentration measurements should be considered as a means to quantify the effects of dilution.

C. Organized Convective Systems

Strong cumulus convection and organized convective systems can transport pollutants from the mixed layer to higher altitudes in the free atmosphere above. Under conditions of strong wind shear these pollutants can then be transported horizontally within the free atmosphere over regional-scale distances and can represent the source of a deposition episode at a remote location. Therefore, it is important to examine the vertical transport of deep cumulus convection as it impacts regional air quality.

Specific questions to be addressed:

What is the vertical distribution of pollutant concentration in the free atmosphere after it has been transported by deep convection?

What is the relationship between the mass of pollutant transported and the intensity of the convection?

D. Tracers for Organized Convective Systems

The objectives of the proposed investigations of transport by organized convective systems identify the need for several conservative gaseous tracers. Releases should be made by aircraft in puff- and/or line-source dimensions. Real-time analysis aboard aircraft is required with time resolution of a few seconds and a secondary need for grab-sample collection.

The currently available suite of perfluorocarbon tracers is ideal for these investigations and poses no serious engineering problems in release logistics. However, fast real-time analysis is not currently possible, which highlights a top priority in analytical research and development. The experimental plan must clearly address the protocol of tracer(s) injection(s); aircraft use especially as it pertains to cross contamination between release and sampling aircraft, meteorological support measurements (Doppler radar); etc.

Although current tracer technology allows for progress in studies of transport by organized convective systems, we make a strong recommendation for development of a remote tracer sensing system. This would greatly enhance the scope of transport studies, simplify them, and greatly reduce their cost.

V. HAZARDOUS MATERIAL DISPERSION

Acute and long-term air quality problems arise from inadvertent releases of hazardous materials to the environment. Examples of such releases are spills (the train derailment of arsenic at Elkhart, Wisconsin); waste disposal (dioxin application to roads in Times Beach, Missouri); fires or explosions (dioxin deposition in a chemical plant fire in Seveso, Italy); and long-term deposition (asbestos deposition from a mining operation at Globe, Arizona, and arsenic deposition from the Asarco smelter in Tacoma, Washington).

Physical properties of the pollutant material are highly relevant to the environmental consequences. Gaseous pollutants, when released in an episode, result in a puff of pollutant, whereas a continuous leak results in a point source. Liquid pollutants lead to mostly hydrological problems unless the liquid is highly volatile (for example, ammonia), in which case it becomes a gaseous pollutant or dries on the surface to become highly divided dry particles that may be suspended by the wind. In this last instance, the liquid-pollutant spill leads to a contaminated aerosol problem. For solid materials, particle size and density are important factors in determining how much of the material may be carried as contaminated aerosol after it is released into the air and how much may become an aerosol after deposition on the surface.

The hazardous material's chemical properties are also important in determining environmental impact. Solubility of the liquid or solid pollutant is important in determining the time scale of the air contamination problem if rainfall is abundant. The pollutant compound's stability is likewise important in determining the time scale of the problem. It's toxicity determines the lower limit of concentration at which the pollutant should be followed (that is, the spatial scale). The reaction of the pollutant with the soil is an important factor that determines whether the pollutant, once deposited on the soil, will be resuspended by winds, and, if it is resuspended, what its particle size distribution will be.

Meteorological (and micrometeorological) influences on the hazardous material dispersion include those of all parameters that determine diffusion in the atmosphere. When emission is limited and stability, toxicity, and solubility are sufficiently low, the deposition problem is probably restricted to local distance scales; when the material is highly stable, highly toxic, and insoluble (for example, plutonium), air transport and dispersion time scales become long, as do distance scales. For pollutants having such long time scales, the effects of the pollutant's resuspension from the soil must be considered. High wind speeds are required to generate appreciable concentrations of contaminated aerosol from contaminated soils because a threshold wind speed must be exceeded. Rainfall greatly influences this resuspension of pollutants from the soil by transporting particles within the soil, increasing the threshold wind speed, and promoting crust formation when wet soil dries.

The tracer used to study the effects of a hazardous material release should have physical and chemical properties similar to those of the material in question. For stable gaseous emissions, for example, a heavy gas tracer could be used. However, the tracer should be nontoxic so that field studies will not produce harmful side effects. Solid particulate tracers should be chosen so as to mimic physical and chemical properties of the pollutant as nearly as possible. The panel suggested the use of nontoxic tracers <u>within the soil</u> to investigate the pollutant movement within the soil and its interaction with soil material (both organic and mineral). Release of the tracer should simulate the conditions of the pollutant's release, as related to spilling and the subsequent aerosol deposition. Sampling and analysis depends on the choice of tracer. Sampling tracers in soil should be done at several depths and over several surface locations to avoid local anomalies. Because most of the hazardous material dispersion will probably be limited to local scales, tracers to study the dispersion should be particularly useful when difficulties such as complex terrain, forest canopy, and land/water interface introduce special diffusion problems.

When designing experiments for dispersion of hazardous materials, we should assign a high priority to classification of hazardous materials, indicating those that have time scales long enough (as determined by solubility, chemical stability, reactivity with soil materials, and frequency of rainfall, as well as toxicity of the pollutant) to require long-term studies of pollutant movement in the soil. This classification project could begin with the list of hazardous substances in the Federal Register and published chemical properties of such materials. When important chemical properties are unknown, experiments should be designed to determine them. This list, including chemical and physical properties of the pollutants, would provide selection criteria for the tracers to be used in soil-pollutant experiments. Soils used in the experiment should be representative of a wide variety in mineralogical, textural, and organic properties. Wetting of the soils should be done at differing rates, and observation of surface crusting and surface pollutant concentrations should be carried out over long periods.

Resuspension studies should be done to relate surface soil tracer concentration and size distribution of dry soil aggregates with physical measurements of wind stress, kinetic energy of saltatory particles, and suspension of tracer material at wind speeds below the saltation threshold wind speed. Additionally, we should study particle suspension, including nonsteady-state erosion, effect of tracer deposit length along wind direction, and convective phenomena such as dust devils.

VI. USE OF SPACE TECHNOLOGY

An enormous technology has been developed over the past few years in the U.S. space program. It is past time that the atmospheric tracer community began to identify areas where this technology can be used. In particular, one thinks of space stations such as satellites, the space shuttle, and eventually, the space laboratory as observation/measurement platforms for boundary layer/tropospheric tracer experiments.

At present, visible plumes from sources of opportunity (fires, industrial plants, dust, volcanoes) and intentional releases could be photographed from space orbit and analyzed by the opacity method or other techniques. We already have the components required to track transponders carrying tetroons from space. At least in principle, this would allow a large number of transponder-coded tetroons to be tracked to global scales and beyond to multi-global scales (several circumnavigations of the globe). Further, it is conceivable that, with the cooperation of atmospheric, space, chemical, and remote-sensing programs, a gaseous tracer system could be developed and plume concentrations could be mapped from space. All these schemes have the advantage of a single data-collection platform whose data are transmitted to a central earth location in real or nearly real time.

To take advantage of this existing technology, such as "star wars" concepts should be presented to the various scientific communities involved and a good coordination program developed. Although tracer specialists are deploying conventional balloons and aircraft platforms in tracer studies that are required now, we should remember that a high payoff for the future can result from pursuing some of the ideas discussed above. Surely, during the initial stages of thinking this subject through, many more useful ideas will evolve.

VII. IN SITU TRACER PRODUCTION

Another possibility for future tracer research is the development of an _in situ_ tracer producer, that is, a device that could take in ambient air and emit it with one or more elements tagged in some way.

(The panel was not sure of any details, but the concept seemed a likely area for consideration.)

NATURAL SOURCES OF INORGANIC AND ORGANIC GASES AND AEROSOLS

I. INTRODUCTION

It is clear that the interactions of inorganic and organic gases and aerosols play important roles in determining the physical and chemical nature of our atmosphere. In addition, it is now recognized that organic and inorganic chemistry are coupled and should be treated as an integrated system. Likewise, the gaseous, aerosol, and aqueous phases must all be considered when evaluating air pollutant impacts.

Both natural and anthropogenic sources of organics (including sulfur-, nitrogen-, and halogen-containing compounds), oxides of nitrogen and ammonia, and sulfur compounds in gas and aerosol phases can contribute to the observed composition of acid precipitation, gaseous and aqueous oxidants, degradation of visibility, cooling and/or heating of the troposphere, and other climatic effects.

Once released into the atmosphere, both natural and anthropogenic emissions will undergo similar long-range transport, and chemical and physical transformations; they finally will be removed from the air by wet or dry depositional processes. Natural sources are generally distributed, whereas anthropogenic sources are largely point sources. Their transport properties may be quite different.

This working group focused on evaluating our ability to differentiate "by the use of tracers" natural from anthropogenic sources of inorganics, organics, oxides of nitrogen and ammonia, and sulfur compounds in gaseous and aerosol form. Natural sources of the materials that were considered included:

- volcanic activity,
- atmospheric electrical activity,
- ;tratopheric injection,
- forest fires (including wood and agricultural burning),
- emissions from vegetation, soils, and biota (land),

87

- soil erosion and suspension, and
- marine sources.

In the past, in the absence of an ability to adequately tag each of these natural sources, it was difficult, if not impossible, to accurately estimate natural contributions to observed levels of acid deposition or visibility degradation.

With the aid of recent advances in tracer technology, reasonable estimates of the natural contribution to these effects may be feasible in the near future.

The principal focus was chosen with regard to the sources' potential contribution to acid precipitation, with the understanding that many other materials are co-deposited with the acids (that is, wet oxidants, aerosols, soluble organics, etc.).

These sources were first examined with regard to their importance in contributing organics, oxides of nitrogen, sulfur (ultimately SO_2 and acidic sulfate), and alkalinity (ammonia, basic soils). Established tracers, potential tracers, or lack thereof for the various sources were then determined. Use of known diurnal and seasonal variability, as well as identification of specific sites where these sources were likely to be important was also discussed and evaluated. The group also deliberated on the adequacy or inadequacy of existing instrumentation for measuring the identified or proposed tracers.

II. GENERAL RECOMMENDATIONS

The group concluded that the following recommendations are appropriate. No specific priority is intended in the order of the statements.

- Because of the complexity of the natural vs anthropogenic source question, naturally existing tracers should be used when possible in conjunction with already developed conservative tracers (for example, SF_6, perfluorocarbons, etc.).

- In almost all cases, statistical methods of analysis should be used to evaluate multiple-component source terms.

- Evaluation of natural sources will require strong interactions between atmospheric scientists, chemists, geochemists, and plant and soil scientists to establish the best time and place for source evaluation as well as identification of new tracers where necessary.

- Source variability should be examined as a function of humidity, region, and season.

- Consideration should be given to the establishment of and long-term commitment to regional background stations and the continuity of existing ones to enhance the data base required for evaluating specific natural source contributions. This will include vertical profiling.

- Improved instrumentation should be pursued in many areas related to the use of tracers, including sampling and analysis.

- Particle tracer development should concentrate on the small-particle fractions.

III. SPECIFIC RECOMMENDATIONS BY SOURCE

A. Emissions From Vegetation, Soils, and Biota (Land)

The group felt this area is one of the most potentially important sources of organics, reduced sulfur gases, and ammonia.

1. Vegetation. Based on previous work, forests were chosen the dominant source of hydrocarbons: deciduous forests releasing principally isoprene, and conifer forests releasing principally alphapinene. Because of the high reactivity of these "terpene" emissions, it was suggested that more stable products might be more readily traced; that is, H_2CO/CH_3CHO ratios

and methacrolein could be useful indicators of isoprene contributions, whereas alphapinonic acid or other optically active products of alphapinene might be used to trace conifer forest input. Use of carbon isotopes (^{13}C and ^{14}C) on specific oxidation products (for example, formaldehyde) was also suggested, which would require that the sensitivity of ^{14}C methods be much improved. Sites for deciduous forests should be chosen in the eastern United States, and likely sites for conifer studies should be selected in the western United States. To evaluate temperature and humidity effects on emissions from these forest types, monitoring on a north-to-south bias in these two areas would be very useful.

Crop emissions should also be examined, but they were given a lower priority for two reasons. First, forests are a larger biomass, and second, it was felt that with the apparent decline of forests in the U.S. and Europe, they should be given first priority.

2. Biota and soil (land). This was considered a potentially important source of hydrocarbons, reduced sulfur gases (eventually SO_2 and acidic sulfate), and ammonia. Based on previous work, wetlands were expected to be the dominant source areas. Potential tracers are CS_2, COS, CH_4, alkanes, ethylene, carbon isotopes (^{13}C and ^{14}C), and sulfur isotopes. Tracer identification and instrumentation development are needed. The Gulf Coast area appears to be a likely candidate for study.

B. Forest Fires (Including Wood and Agricultural Burning)

The group felt that fire is an increasingly important source of organic gases and aerosols, NO_x, and--to a small extent--SO_2 and sulfate. Potential tracers include potassium in submicron aerosols, methylchloride, alkylated polycyclic aromatic hydrocarbons, vanillin and other biomarkers, and carbon isotope (^{13}C and ^{14}C) determinations on elemental carbon. Improved sensitivity for ^{14}C determinations would be very desirable. Agricultural burning will be strongly regional. Wood burning is more widespread and might be evaluated best in the northwestern United States. Large forest fires could be a large-event source. Alaskan and/or Canadian fires would be likely candidates for study.

C. Soil Erosion and Suspension

This is major source of alkalinity for neutralizing acidic species and precipitation. Tracers include trace metals and minerals, sulfate, nitrate, radon, and thoron. The group decided that it is particularly important to determine the particle size vs trace metal concentrations, particularly in the 0.1-to 1-µm range. Improved instrumentation may be required. Trace metal variability as a function of particle size appears to be the key to soil source differentiation because characterization as a function of particle size can be used to improve statistical analysis; in addition, existing data support the theory that size fractionation does occur. Areas to study include the Great · Basin area, the south-western United States, and midwestern croplands where suspension by wind and mechanical processes occurs frequently.

It is important to understand that the compositions of bulk soils are not ideal for characterizing sources because fractionation is very likely to occur during the mobilization process. It is the composition of the mobilized aerosol that we should measure. Improved instrumentation may be needed to determine the size-composition distribution, particularly in the 0.1-to 1-µm range.

Studies of particle growth dependence on relative humidity and rainout are needed to estimate removal rates and balance trace metal source-receptor budgets.

D. Marine Sources

It is well known that the ocean can be the source of particulate matter in the atmosphere (sea salt). Recent data also indicate the ocean can be a source of organics and reduced sulfur gases (for example, dimethyl sulfide). Characteristic tracers of marine air include sulfur, isotopic signature of sulfate, halides (inorganic and organic), especially gaseous iodine species, and low radon content.   With regard to background levels of acidic sulfate in the western United States, dimethyl sulfide has already been indicated as an appreciable contributing factor.   Further work is warranted, particularly from air masses originating in the Gulf of Mexico, the Gulf of California, and the Pacific Ocean.   Better measurement techniques that are selective and sensitive to reduced sulfur (dimethyl sulfide) should be pursued.

It is important to recognize that the natural inorganic and organic gases and aerosols of concern are often quite reactive both chemically and physically and undergo transformations in transport not unlike those of their anthropogenic counterparts.   Because of this source, tracers such as isotopic signatures may be used for specific materials like dimethyl sulfide or oceanic sulfate.

E. Volcanic Activity

Although volcanoes are an important source of sulfur and particulate matter, the group decided that their effects on acid precipitation would be short term (that is, event dominated) and insignificant when compared with other sources.   Recognized as an important factor for global climatic effects, volcanic activity has a lower priority as a source of acidic precipitation.

F. Atmospheric Electrical Activity

Atmospheric electrical activity is a potentially important source of NO_x (NO, NO_2, HNO_3) and ozone.   No specific tracer was identified.   Nitrous oxide (N_2O) could be a potential tracer and should be evaluated.   Because of frequency of occurrence and low-background NO_x levels, the mountainous areas of the southwestern United States are likely sites for studies of this source term.   This area has received considerable attention in previous investigation and therefore, it should be given a lower priority.

G. Stratospheric Injection

This is probably a small source of NO_x.   Tropopause folds may account for significant injections of O_3 that could be evaluated from existing data.   In addition, Be[7] could be used as a tracer for this process.

IV ADDITIONAL GENERAL COMMENTS

It was frequently noted that the processes by which both natural and anthropogenic gases and aerosols are removed are dominated by certain chemical and physical removal processes. For example, hydroxyl radical (OH) is key in the gas-phase oxidation of many compounds, and many pollutants undergo wet and dry deposition in the environment. Other species, such as H_2O_2 and other wet oxidants, may be important in cloud processes. Other processes and reactions are also of potential importance (removal by NO_3 at night, photolysis, reaction with ozone).

For many of these removal and transformation processes, it would be extremely valuable if a suite of nonconservative tracers were developed that, in conjunction with the already available conservative tracers, could measure these mechanisms. These tracers should have low natural backgrounds, low toxicity, and high detection sensitivity. Reactive/ECD sensitive perfluorocarbon (or other halide) materials are one possibility. This approach will require considerable development and possible use of more expensive analysis equipment (for example, negative ion capillary gas chromatography/mass spectrometry).

The importance of measurements for both natural and anthropogenic transformation and removal processes should certainly not be underestimated. Measurement techniques and equipment should be given high priority.

Part IV

Invited Presentations

TRACERS IN TRANSPORT AND DIFFUSION

Sumner Barr

Atmospheric Sciences Group
Earth and Space Sciences Division
Los Alamos National Laboratory
Los Alamos, NM 87544

ABSTRACT

Transport and turbulent diffusion are the traditional tasks to which tracers have been applied in the atmosphere since Richardson's innovative experiment in 1926. There are several primary objectives in tracer experiments from the fundamental goal of relating diffusion to elements of atmospheric dynamics to the practical goals of diffusion and the testing of release mechanisms. Applications of tracer experiments come from many fields, including agriculture, national defense, safety and environment, and transportation.

In this paper we will summarize the set of diffusion trials upon which the conventional parameterizations of turbulent diffusion are based. A quick examination of the limitations of classic diffusion experiments leads us to a survey of experiments in special environments. The special environments include nonideal but nonetheless very practical variations of site structure, meteorology, and source effects. The site factors include complex terrain, shorelines, vegetative canopies, and urban complexes. Very light wind conditions and the convective boundary layer constitute meteorological factors receiving attention. Long-range tracer experiments and descriptions of concentration fluctuations rely on meteorological factors but also include features of the underlying surface. Finally, source factors that have been studied include tall stacks, cooling towers, spray disseminators for agriculture, very dense gases, and a host of natural tracers such as volcanoes, radon gas, resuspended dust, and sea surface emissions.

A number of recent theoretical, organizational, and instrumental developments have a major impact on the role of tracers in atmospheric transport and diffusion. Model validation efforts are forcing us to rethink the questions surrounding air concentrations from a tracer source, including the recognized role of time and space fluctuations. Remote sensing techniques offer an opportunity both to learn more than we could by in situ samples and to understand the relationship between the new measurements and our traditional concepts.

I.   INTRODUCTION

Perhaps the simplest task we can ask of a tracer is to follow the movement of some prototype material through a dynamic system. If that system
is the atmosphere, the motion is three dimensional, nonstationary, and disper-
sive because of turbulence, and it covers very large volumes. The motion we
attempt to document with tracers includes the bulk trasport (horizontally
and vertically) and the dilution associated with turbulent diffusion. Because
the spectrum of atmospheric eddies is continuous and extremely broad-rang-
ing, from the millimeter size of the dissipation range of motions all the way
up to global size circulation features, it is a little artificial to partition the
problem into transport and diffusion. Eddies that play a transport role in the
first few hours during the travel of a puff of material become "diffusers"
when the puff grows large enough at longer travel times. The convenience of
distinguishing between transport and diffusion is reflected in the design and
interpretation of experiments.

Following Draxler,[1] we can identify several rationales for tracer experiments in the atmosphere. According to Draxler, "the purest diffusion experiments attempt to relate the dynamic and physical characteristics of the
earth-atmosphere system to the mean and turbulent atmospheric structure
and, in turn, to relate these characteristics to diffusion." Studies with these
objectives require considerable resources in planning and implementation, and
although they yield the most fundamental and, therefore, general results, it is
often difficult to justify the time and expense when practical applications are
the motivation. A more modest characterization of the atmosphere may rely
on a series of atmospheric states defined by one or more parameters including, for example, temperature lapse rate, wind speed, time of day, cloud
cover, surface roughness, and gross turbulence statistics. Other objectives
include the diffusion climate of a site and the effects of a release mechanism. Another way of viewing the objectives of tracer experiments is to
examine the three major factors that they address: source (altitude, geometry, duration, buoyancy, and particulate content), site (ground condition,
terrain, and vegetation), and meteorology (wind field, turbulence, mixed layer
depth, time variability, and heterogeneity).

There is a wide range of practical applications for tracer experiments.
In many cases the tracer allows us to document a potential airborne hazard
without actually emitting hazardous material or to evaluate the expected

outcome of an expensive process alteration. Some examples of the former
case are testing of some rocket fuels, developing chemical defense strategies, establishing safety procedures for possible accidents in the handling and
transporation of hazardous materials, and assessing the fire and explosion
hazards present in handling special materials such as liquified natural gas.
The practical use of tracers for economic benefit includes designing emissions systems, for both air quality control and dissemination systems such as
in forest and crop spraying or cloud seeding, and siting of industrial facilities.
Other applications include source attribution, which is currently of interest in
the acid rain problem.

In this paper we will briefly review the traditional diffusion experiments and summarize parameters that have evolved from the empirical basis
of these experiments. Generally, they emphasized the meteorology of the
boundary layer, including wind, temperature, and turbulence structure. Complications in the underlying surface were avoided in many of the traditional
experiments. In the 1970s, the realities of plant siting forced a consideration
of more complex surface conditions, and therefore more recent experiments
have addressed complex terrain, shorelines, cities, and forests. These are
surveyed in a section on special environments and recent tracer activities in
long-range transport and the meteorological conditions of very low wind
speeds and the convective boundary layer.

II.   TRADITIONAL DIFFUSION TRIALS

Tracer experiments were conducted in the 1920s, but technology advances in the 1950s and 1960s offered the first real opportunity to perform experiments that were practically designed in terms of travel distance and time
and space-sampling resolution. The first experiments related diffusion to the
physical characterization of the earth/atmosphere system, accounting for
surface roughness, wind, temperature, and turbulence profiles. With the
emphasis on basic atmospheric structure, as many as possible of the complicating meteorological and surface variations were designed out of the experiments; efforts were made to find flat sites of uniform roughness and to run
the experiments during stationary meteorological conditions. Recognition of
the natural variability of the turbulent atmosphere prompted experiments
designed with a large number of realizations (up to 70 per series). Typically,
a detailed sampling grid consisting of crosswind arcs at varying distances

(for example, 0.1 to 10 km) from a well-controlled source formed the basis of
the experiments that followed the concept of an expanding plume of tracer
material from a point source. When feasible, vertical profiles or tracer concentrations were also measured. The Gaussian plume models available for
interpretation then and now contain parameters sensitive to

- source function time, elevation, and geometry (point, line, and
  area);
- meteorological static stability, profiles of wind and temperature,
  and turbulence statistics (assuming stationarity and homogeneity);
  and
- surface roughness and uniformity.

The traditional experiments produced a remarkably useful empirical
basis for estimating atmospheric diffusion. Curves of lateral and vertical
plume growth (as shown in Fig. 1) have been reproduced and used for practical estimations thousands of times. The empirical contributions include not
only the form of growth curves with travel distance, but also the meteorological dependencies guiding the selection of the proper curve from the family
of functions.

The empirically based Gaussian plume concentration estimation methods found a hungry user community after the Clean Air Act was passed in
1970. Environmental impact statements were required for a large number of
potentially air-polluting facilities, and the methods derived empirically from
near-surface sources over flat terrain to 1 km were applied to tall stack
sources over all kinds of terrain to distances of 100 km. It became clear that
a similar approach combining existing theory with a series of tracer experiments was needed for a variety of what Gifford[2] refers to as exceptional
cases. In the next section we will summarize efforts in a number of special
environments that require additional consideration because of source, site, or
meteorology factors influencing the turbulence or transport wind field beyond
the levels considered in the traditional methods.

III. SPECIAL ENVIRONMENTS

Many sites are of great practical interest but are not well characterized by the diffusion formulas developed from traditional experiments. A

major complication is the variability of the underlying terrain; it can influence the transport and diffusion process as little as to cause a nonhomogeneous turbulence field or as much as to totally dominate the wind and turbulence regime that governs tracer dispersal. When the surface is ocean or lake water or a shoreline, another set of meteorological dependencies is operative. Forest and other vegetative canopies affect turbulence, winds, and temperature, dominating the conditions within but also extending their influence to the region above the canopy. Urban complexes alter the wind and temperature fields through the urban heat island. Alterations in turbulent mixing occur from the heat-island mechanism as well as from the roughness effect of many buildings.

The meteorological conditions themselves offer special environments for transport and diffusion. Under very low wind speeds, turbulent meander is not small compared with the mean transport; the whole problem must be treated differently than in the traditional case. Also, the convective boundary layer resisted treatment by conventional methods until Deardorff[3] introduced an additional scale parameter. Consideration of longer range travel, from a hundred to a few thousand kilometers requires the introduction of atmospheric properties on larger time and space scales, including diurnal variations. Also, the larger scales are likely to cover more types of underlying terrain than a small-scale tracer experiment can.

A. Complex Terrain

Experiments in complex terrain have tended to follow two of the main objectives described above. To characterize the atmospheric physics of some terrain interactions, efforts have been made to isolate a given topographic feature such as a simple slope, ridge, or valley. Interest in siting industrial plants within areas of complex topography has prompted quite a few tracer studies to establish the diffusion climate of the particular site.

Experiments aimed at physical-dynamical characterization usually have as much supporting meteorological data and as many realizations as budgets will allow. Because of the well-documented spatial variations of terrain-influenced wind, temperature, and turbulence fields, the typical meteorological arrays tend to be more extensive than those for simple terrain. Since the earlier experiments, the trend is toward deeper profiles and more widespread arrays.

Smith[4] describes an early tracer experiment in which tracer was released from an aircraft upwind of a quasi-linear ridge. The mechanisms of interest were lee-wave phenomena that deposit the elevated released tracer on the lee side of the ridge. He observed several cases of lee-side deposition that were moderately well predicted by the mechanism he proposed (Fig. 2). In a very si nple study, Gedayloo et al.[5] observed deposition caused primarily by viscous flow separation from a sharp-edged ridge (Fig. 3). They used a power plant stack as a tracer of opportunity and deployed a long-term monitoring array. Super[6] describes a cross-ridge tracer study conducted in preparation for a cloud-seeding project. Rowe[7] released neutral balloons in a cross-ridge wind in an attempt to document buoyancy wave properties.

More recently, to document plume impingement characteristics, the US EPA has supported tracer studies of stable flows approaching a ridge and an isolated hill. A mechanism of primary concern is the tendency of a stable flow to partition into a lower segment that flows around (or along) an obstacle and an upper portion that flows over the obstacle. The dividing streamline that separates the two appears to be reasonably well predicted by potential vs kinetic energy criteria.

The valley is a major topographic structure studied with tracers. Valleys tend to be the most hospitable sites in complex terrain for habitation and industrial siting; therefore, they are common sites for environmental assessments. Quite a few documented physical processes occur in valleys, so they are a good site at which to study certain meteorological fundamentals. The DOE ASCOT program has focused on nocturnal valley phenomena that influence transport and dilution processes. The drainage layer of cool air that sinks down the valley sidewalls and then flows out of the valley along its major axis seems to be somewhat decoupled from the air above and also appears to have a characteristic turbulence pattern. We hope that if diffusion in the drainage domain is characterized, the procedures for estimating this diffusion will be broadly applicable because of the high reproducibility of the locally driven wind domain. ASCOT has performed tracer experiments in two valleys in California and one in Colorado. The same personnel gained related experience on other missions that included other valleys. Barr[8] has found that several mechanisms must be considered when interpreting the concentration patterns from point sources in a valley. First, the lateral and vertical growth rates appear to be well predicted by representative turbulence measurements but not so well handled by the more highly parameterized estimation schemes that are based on flat-terrain data. Second, quasi-stagnant pooling of cool air in poorly draining areas of valleys affects the area and duration of exposure to airborne materials. Third, sidewalls tend to constrain the lateral growth, but the top of the drainage layer does not seem to bound vertical growth. Mean vertical motions caused by horizontal convergence produce deeper and more complex tracer plumes than would be expected over flat terrain. Other nighttime valley tracer studies have been reported by Start et al.[9] and Wolf et al.[10] In general, diffusion was more vigorous than predicted by conventional flat-terrain parameters. Suggested mechanisms include enhanced roughness and lateral gravity currents from sidewall corrugations.

Several rough-terrain tracer experiments have been conducted over an ensemble of features. The interpretation is somewhat complicated by interactive effects from more than one terrain type, but observations tend to suggest several common features. Hinds et al.[11] found situations where tracer seems to skip over shallow valleys and other instances where it sweeps down through the valleys.

B. Vegetative Canopies

The prevalence of vegetative cover over major portions of the earth's surface indicates the importance of understanding turbulent exchange processes within and above vegetative canopies. Although in some cases the canopy volume itself is the total domain for diffusion (for example, forest and crop spraying or defense against chemical weapons), more often it represents the source, deposition, or receptor phase of a more general atmospheric transport situation. The frictional and radiative energy balance roles of the vegetative canopy immensely alter the turbulence and wind structure. Available surface area, deposition, and biological uptake of particles and gases are the dominant and currently very poorly understood processes.

Tracer experiments have involved three major regions associated with canopies, each characterized by different physical processes. Tracing a plume from a finite source will usually involve one or more of these canopy domains. The three zones discussed here are the regions within the canopy, above the roughness elements, and at the leading edge of a canopy.

The main characteristics of the region within the continuous canopy are a major reduction in mean wind, extensive turbulence generated by the roughness elements, an alteration of stability caused by radiative effects at the top of the canopy, and enhanced deposition resulting from a large available surface area. Several empirical studies have established the basic structure of turbulence within crop and forest canopies. There are good indications that these turbulence features can be generalized and incorporated into existing computational diffusion methods, as discussed by Barr and Clements.[12] In the range of travel times from 10 seconds to 20 minutes, the calculated plume-spreading rate ($\sigma_y \propto t^{0.7}$) is slower than in an open boundary layer. This is consistent with the observations of Raynor et al.,[13] who found relatively slow lateral dispersion of tracer aerosol within canopies. Draxler[1] shows a composite diagram of four series of experiments that show evidence of reduced growth rate for individual series, although gross dimensions remain consistent with the general formulas.

Whereas in the open boundary layer lateral dispersion is more effective than vertical spread, especially in stable conditions, the roughness elements in the forest tend to enhance the vertical component of turbulence. The intensity and scale of vertical turbulent fluctuations tend to be comparable with those of the horizontal eddies, and plume growth follows the same pattern.

Several authors (Baynton et al.[14] and Petit et al.[15]) have addressed the problem of pollutant sources of various geometries above a quasi-uniform vegetative canopy. The problems have been handled in three parts: (1) diffusion in the open atmosphere above the canopy, (2) interchange between the below-canopy and above-canopy zones, and (3) dispersion within the roughness elements.

The open-air component may be handled by conventional methods for the source and meteorology, and the subcanopy dispersion by the methods described above. The interchange has been addressed by Baynton et al.[14] through the concept of penetration ratio, a ratio of time-integrated concentrations within the canopy and overhead. The values of the ratio were related to lateral turbulence intensity through a regression analysis, giving a predictive formula for exchange into one particular jungle canopy. Empirical determinations of a similar parameter for a wide range of canopy types is needed to broaden the basis for estimating exchange rates.

In the case of more slowly varying concentration fields, the flux of pollutant at the top of the vegetation may be evaluated through flux-gradient relationships with observed concentration gradients and various methods for estimating diffusivity.

Few applications deal with uniform canopies of infinite extent. In practical cases, we must account for phenomena near the edge. of forests. Rapid deceleration of the mean flow by trees tends to produce extensive turbulence and forces air flow up over the forest at a leading edge. Both these processes contribute to rapid dispersal of pollutant that enters a wooded area from the windward side. After a transition region of about 20 to 30 tree heights, the diffusion becomes characteristic of the uniform canopy discussed above.

C. Over Water and Shorelines

Tracers have been quite useful in dispersion problems over water and shorelines, areas for which there are many practical requirements. Experiments conducted on the east, west, and Gulf coasts of the US and the Great Lakes demonstrate some common characteristics of this domain. Several dominant physical processes are seen over shorelines that are not generally present at sites without the water/land discontinuity. One is the thermal internal boundary layer (TIBL) created by differential heating of the land and water surfaces. Tracer plume behavior in the stable layer outside the TIBL is often very different from the convective mixing within it. A common occurrence is the fumigation of a stable elevated stack plume when it intersects the TIBL. The mean transport velocity field is affected by the diurnally and spatially varying sea breeze. In addition to the different stability environment over land and water, the aerodynamic smoothness of the water surface helps account for a much lower level of turbulent energy and, therefore, slower diffusion of tracer over the water than over the land. Minott and Shearer[16] summarize a series of tracer experiments in which larger scale vertical velocity fluctuations play a role. These fluctuations were accounted for by partitioning the total dispersion into relative diffusion and meander components. Several mechanisms contribute to the larger eddies. Convective eddies advected from over water as well as those generated within the TIBL exhibit persistent lifetimes. Also, a deceleration of the mean onshore wind as it encounters a rougher surface at the land/water interface creates a divergence of streamlines in the vertical plane. This vertical impulse may result in a series of buoyancy waves. Also, the terrain surface may initiate similar phenomena. Minott and Shearer find that beyond about 100 m down-wind the meandering component dominated the total vertical diffusion.

A great deal has been learned about shoreline effects from observing and monitoring the stack plumes of power plants located at shorelines. Lyons[17] shows evidence of dramatic changes in the behavior of a single plume as it intersects the boundary of the TIBL (Fig. 4). The fanning charac-ter of the elevated plume in the stable, smooth over-water environment fum-igates quickly to the ground at the inland location where the convective TIBL has grown to intersect the plume.

D. The Unstable Planetary Boundary Layer

The atmospheric boundary layer under strong insolation and light-to-moderate wind speeds exhibits unstable thermal stratification and associated convection currents. An easily observed consequence of the convection is the increase in the magnitude of wind fluctuations near the ground. Also, dilu-tion of effluent plumes proceeds much more rapidly in unstable conditions than in other conditions. When considering diffusion in unstable conditions, it is necessary to view the entire unstable boundary layer and its associated overlying stable layer as a system, frequently referred to as the mixed layer. Recent studies of the dynamics and structure of the unstable boundary layer by Deardorff,[3] Wyngaard and Cote,[18] Clarke et al.,[19] Yamada and Mellor,[20] and others have provided the basis for a comprehensive picture of the time and space evolution of the mixed layer.

The important features of the mixed layer for the purposes of diffusion estimation are

(1)   the depth of the layer,

(2)   the magnitude of vertical and lateral turbulent fluctuations in the layer, and

(3)   the profile of turbulence with height.

Scaling studies treat the depth of the convective mixed layer z_i as a funda-mental quantity. It depends on time of day, intensity of surface heat flux, and thermal stability of the atmosphere outside the layer.

Briefly, the mixed layer is driven by heating at the ground surface, which produces unstable temperature stratification in the lowest layers and allows vigorous convection to occur. The convection promotes very efficient flux of virtually every transportable quantity in the vicinity of this surface layer. Convective plumes then traverse the quasi-adiabatically stratified interior of the mixed layer quite efficiently. At the top of the layer, the plumes encounter an interface with more stably stratified air. The processes that take place at the interface include a penetration of the stable air by an inertial overshoot and an exchange of energy from relatively large vertical convective scales to smaller scales of turbulence in all directional components.

One of the properties of pollution transport in the mixed layer appears to be a tendency to concentrate material into a secondary maximum near the top of the layer at relatively short travel distance (for example, Deardorff and Willis[21]). Further downstream the material spreads from the upper and lower boundaries of the mixed layer toward the center, producing a uniform concentration profile in the vertical direction.

Some recent tracer experiments conducted in the unstable boundary layer include those of the EPRI plume validation study reported by Bowne[22] and the monitoring of smelter plumes by Carras and Williams.[23] The close-in concentration patterns are widely distributed and very nonuniform--not representative of the classical plume picture.

E. Stagnation Environment

The condition of very low mean wind speed and stable temperature gradient represents one of the poorest conditions for dilution of atmospheric emissions. The problem is compounded by the inapplicability of traditional modeling techniques under these limiting conditions. Although an extension of conventional diffusion models indicates diminished diffusivity with ever-increasing thermal stability, observations show that beyond a threshold stability, the average spread of material actually increases. The particular wind speed--stability levels at which the minimum dilution rates occur and the character of the diffusivity function in the vicinity of the minimum are site dependent and probably depend on wind direction at a given site. The cause of the enhanced dilution under low wind speed, stable conditions is the presence of a low-frequency meander of wind direction. The vertical component of turbulence σ_w appears to continue to diminish under the influence of thermal stability while the low frequency fluctuations keep σ_θ above a site-dependent threshold level.

The practical need for study of the low wind speed, stable diffusion characteristics is by no means trivial. For significant fraction of the time at most locations, winds are less than 2 m/s. Furthermore, they generally represent the worst cases of diffusion for a particular facility; hence, the estimates contribute heavily to design concepts. If generalized estimation methods routinely yield overconservative results, expensive over-design may result. However, lack of conservation may result in damage to the environment. The answer is an adequate understanding of this condition to produce reasonable estimates of the dilution of atmospheric emissions.

F. Urban Areas

There are a number of aerodynamic, thermodynamic, and chemical factors that characterize the urban dispersion problem. One of the most significant factors is the complexity of the source configuration. Cities are affected by isolated point sources at ground level and at elevated levels. They are superimposed on quasi-uniform, low-level residential sources and localized intense area sources associated with industrial activity. Added to these are an array of line sources produced by time-dependent traffic flow and the background of material transported from nearby cities. Of course, for particular pollutants one or more of the source types may be dominant, but the list of potential sources serves as an example of the need for high-quality source inventories.

In addition to the increased complexity of source configuration, the urban complex alters the basic meteorological structure governing transport and diffusion, including mixing depth, mean wind direction and speed, temperature structure, and related turbulence. Also, alternated mechanisms of chemical and physical transformation and wet and dry deposition are fundamental characteristics of the urban problem.

It has been well established that cities have surface temperatures several degrees warmer than the surrounding countryside. Although the mechanics of the heat island is a challenging problem in micrometeorology, the consequences of the phenomenon are of primary interest in the pollutant transport problem. One such consequence is an airflow circulation created by low-level confluence of cool, clean, nonurban air to replace the rising heated air over the city. Superimposed on a moderately weak external wind, the circulation tends to tilt downwind with altitude, giving patterns of alternating subsidence and ascent. The motion and temperature fields are closely interrelated and produce regions of stability and instability. Figure 5 from Clarke[24] shows an along-wind cross section of stable layers as well as unstable air over Cincinnati.

The METROMEX Program was a cooperative multiagency field program aimed primarily at investigating inadvertant weather modification produced by the St. Louis urban complex. Some of the results are directly applicable to transport and dispersion, including wind fields (Ackerman)[25] and modification of the capping inversion, which delineates the mixed layer (Spangler and Dirks).[26] Also, of course, there is evidence that the combined effects of particulates as cloud condensation nuclei, altered stability, and vertical velocities may produce a suggested increase in the precipitation. This, in turn, serves to alter the scavenging of airborne pollutants. The interactions are very complicated and the role of these mechanisms as a sink for urban pollutants has not been quantified.

One other fundamental feature of cities that should be singled out is roughness in the scale of building sizes. This variation will increase friction and retard mean wind flow; it will also increase turbulence levels and affect plume growth parameters.

G. Long-Range Transport

The recent development of highly sensitive tracers has allowed us to consider tracer experiments in the range of transport distance from a hundred to several thousand kilometers. We are now able to address many practical problems on regional to continental scales. A series of demonstrations of the perfluorocarbon and heavy methane tracer systems have been reported by Cowan et al.,[27] Fierber et al.,[28] and Fowler and Barr.[29] In fall 1983, the CAPTEX program produced a combined tracer distribution and meteorological data set that, when analyzed, should help establish a foundation for estimating transport and dispersion on the 1000-km scale. Some analyses of tracers of opportunity have been carried out by Heffter,[30] Gifford,[31] Barr,[32] Walton,[33] Randerson,[34] and Carras and Williams.[35] An evolving theoretical framework discussed by Gifford[36] and including mesoscale dynamical concepts from Larsen et al.,[37] Gage,[38] and Lilly,[39] has a great deal of promise for successful explanation of long-range dispersion.

IV.   DISCUSSION

The review of diffusion in special environments discussed above points out some important facts about our understanding of the processes and, subsequently, our ability to accurately estimate the diffusion. Diffusion over simple plane surfaces depends on a basic concept of energetics in which kinetic energy is transported downward through a boundary layer from the mean flow above and is dissipated by friction near the surface. The important profiles are governed by such features as the surface roughness, the total energy available, and the static stability. In these environments there are a number of complications, including internal sources and sinks of turbulent energy, locally generated circulations, and horizontal inhomogeneities in the mean and turbulent wind and temperature fields. Systematic mean vertical velocities affect the behavior of tracer plumes. Also, whereas the spectral exchange of turbulent energy in the simple case is usually assumed to be a downscale cascade with a well-defined functional form, energy sources in special environments are often in several scale ranges, which gives rise to multiscalar exchange as well as space-time variabilities. In some cases, especially in regional-scale transport, this leads to a basic ambiguity between the mean wind and turbulence.

In the preceding section, we described some examples of physical phenomena that are present in selected special environments and significantly influence the patterns of tracer concentration. The examples compose nowhere near an exhaustive list of the observed phenomena, but are offered to show the kinds of complications that attend tracer dispersion in most real sites of interest. Dispersion estimates made without considering the complicating physics can be seriously in error. A series of tracer experiments designed and analyzed flexibly enough to recognize the inherent aerodynamic and thermodynamic "surprises" should be performed in special environments.

V.   SUMMARY AND CONCLUSIONS

In this paper, we have explored several rationales for performing tracer experiments--from exploration of basic atmospheric dynamics to practical simulation of pollution sources. We reviewed the traditional diffusion experiments and their extension to an assortment of special cases where elements of atmospheric physics complicate the problems.

There are a wide variety of gaseous and particulate tracers available today to perform many important types of transport and diffusion experiments from a few hundred meters to thousands of kilometers. There are still many atmospheric problems that require tracer experiments to obtain reliable answers. These include different conditions of meteorology, many types of site, and a range of time and space scales.

We recommend a recursive process in applying tracer experiments to practical problems. The steps of the process should be to

(1)   examine the governing physics of the problem.

(2)   factor the best of what is known of the problem into the design of the tracer and supporting meteorological measurements.

(3)   perform the experiments.

(4)   aggresively search the experimental results for shortcomings and identify the physical reasons. (This step is most often skipped, but it may be the critical one to really extend our knowledge. Step 4 leads us back to Step 1.)

(5)   repeat the process.

Tracer experiments are a critical tool in studying atmospheric turbulence and diffusion; the resources for using them on the widest variety of practical problems have never been better. We should continue our efforts to develop tracer technology. At present, when the concentrations are low, we must collect samples and transport them to the laboratory for analysis; the value of real-time analysis of concentrations at one-part-per-trillion level and lower is enormous. Also, remote sensing tools such as lidar should continue to be exploited because they can provide fast response and spatial coverage that are unattainable by in situ sampling.

REFERENCES

1. Draxler, R. R., 1981, Diffusion and Transport Experiments, Chapter 8, in Atmospheric Science and Power Production, U.S. DOE/TIC-207601.

2. Gifford, F. A., 1976, Turbulent Diffusion-Typing Schemes: A Review, Nucl. Safety, 17:68-86.

3. Deardorff, J. W., 1972, Numerical Investigation of Neutral and Unstable Planetary Boundary Layers, J. Atmos. Sci., 29:91-115.

4. Smith, T. B., 1965, Diffusion Study in Complex Mountainous Terrain, U.S. Army Chemical Corps., Dugway Proving Ground, Utah, Meteorology Research Inc. Report MRI 65 FR-236, Altadena, California.

5. Gedayloo, T., S. Barr, W. E. Clements, and L. E. Wangen, 1979, Behavior of a Tall Stack Plume in Flow Over a Ridge, Los Alamos National Laboratory Report LA-7632-MS, 15 pp.

6. Super, A. B., 1974, Silver Iodide Plume Characteristics Over the Bridger Mountain Range, Montana, J. Appl. Meteorol., 13:62-70.

7. Rowe, R. D., 1980: A Simple Two-Layer Model for Stable Air Flow Across Terrain Features. Proceedings of the AMS-APCA 2nd Joint Conference on Applications of Air Pollution Meteorology, New Orleans, pp. 559-566.

8. Barr, S., 1982, A Comparison of Lateral and Vertical Diffusion in Several Valleys, LA-UR-82-3297, Proc. Sixth Symposium on Turbulence and Diffusion, American Meteorological Society, Boston, Massachusetts.

9. Start, G. E., C. R. Dickson, and L. L. Wendell, 1975, Diffusion in a Canyon Within Rough Mountainous Terrain, J. Appl. Meteorol., 14:333-346.

10. Wolf, M. A., ed., 1972, Parachute Creek Valley Diffusion Experiments. Pacific Northwest Laboratory Report to Colony Development Operation, Sept. 1972, 25 pp.

11. Hinds, W. T., 1970, Diffusion Over Coastal Mountains of Southern California, Atmos. Environ., 4:107-124.

12. Barr, S. and W. Clements, 1981, An Overview of Diffusion Modeling: Principles of Application, Chapter 13, in Atmospheric Science and Power Production, U.S. DOE/TIC-27601.

13.   Raynor, G. S., J. V. Hayes, and E. C. Ogden, 1974a, Particulate Dispersion Into and Within a Forest, Boundary Layer Meteorol., 7:429-456.

14.   Baynton, H. W., W. G. Biggs, H. L. Hamilton, P. E. Sherr, and J. B. Worth, 1965, Wind Structure In and Above a Tropical Forest, J. Appl. Meteorol., 4:670-675.

15.   Petit, C., M. Trinite, and P. Valentine, 1976, Study of Turbulent Diffusion Above and Within a Forest-Application in the Case of SO_2, Atmos. Environ., 10:1057-1063.

16.   Minott, D. H. and D. L. Shearer, 1977, Development of Vertical Dispersion Coefficients for a Shoreline Environment, The Research Corp. of New England, Report C00-4026-2, 52 pp.

17.   Lyons, W. A., 1975, Turbulent Diffusion and Pollutant Transport in Shoreline Environments, in Lectures on Air Pollution and Environmental Impact Analyses, American Meteorological Society, Sept. 29-Oct. 3, 1975, Boston, Massachusetts, pp. 136-208.

18.   Wyngaard, J. C. and O. R. Cote, 1974, The Evolution of a Convective Planetary Boundary Layer - a Higher-Order Closure Model Study, Boundary Layer Meteorol., 7:298-302.

19.   Clarke, R. H., A. J. Dyer, R. R. Brook, D. G. Reid, and A. J. Troup, 1971, The Wangara Experiment: Boundary Layer Data, Tech. Paper 19, Division Meteorological Physics, SCIRO, Aspendale, Victoria, Australia.

20.   Yamada, T. and G. Mellor, 1975, A Simulation of the Wangara Atmospheric Boundary Layer Data, J. Atmos. Sci., 32:2309-2329.

21.   Deardorff, J. W. and G. E. Willis, 1974, Physical Modeling of Diffusion in the Mixed Layer, Symposium on Atmospheric Diffusion and Air Pollution, American Meteorological Society, Sept. 9-13, 1974, Boston, Massachusetts.

22.   Bowne, N. E., R. J. Londergan, and M. K. Lin, 1982, Plume Model Validation Results, Proc. Sixth Symposium on Turbulence and Diffusion, American Meteorological Society, Boston, Massachusetts.

23.   Carras, J. N. and D. J. Williams, 1982, Observations of Vertical Plume Dispersion in the Convective Boundary Layer, Proc. Sixth Symposium on Turbulence and Diffusion, American Meteorological Society, Boston, Massachusetts.

24. Clarke, J. F., 1969, Nocturnal Urban Boundary Layer over Cincinnati, Ohio, Mon. Weather Rev., 97:582-589.

25. Ackerman, B., 1974, METROMEX: Wind Field over St. Louis in Undisturbed Weather, Bull. Am. Meteorol. Soc., 55:93-95.

26. Spangler, T. C. and R. A. Dirks, 1974, Meso-Scale Variations of the Urban Mixing Height, Boundary Layer Meteorol., 6:423-441.

27. Cowan, G. A., D. G. Ott, A. Turkevich, L. Machta, G. J. Ferber, and N. R. Daly, 1976, Heavy Methanes as Atmospheric Tracers, Science, 191, 1048-1050. Also Los Alamos National Laboratory Report, LA-5936-MS.

28. Ferber, G. J., K. Telegadas, J. L. Heffter, C. R. Dickson, R. N. Dietz, and P. W. Krey, 1981, Demonstration of a Long-Range Atmospheric Tracer System Using Perfluorocarbons, NOAA Tech. memo ERL ARL-101.

29. Fowler, M. M. and S. Barr, 1983, A Long-Range Atmospheric Tracer Field Test, Atmos. Environ., 17:1677.

30. Heffter, J. L., 1965, The Variation of Horizontal Diffusion Parameters with Time for Travel Periods of One Hour or Longer, J. Appl. Meteorol., 4:153.

31. Gifford, F. A., 1982, Horizontal Diffusion in the Atmosphere: A Lagrangian-Dynamical Theory, Atmos. Environ., 16:505-511.

32. Barr, S., 1983, The Random Force Theory Applied to Regional Scale Tropospheric Diffusion, draft paper, 7pp.

33. Walton, J. J. 1974, Scale-Dependent Diffusion, J. Atmos. Sci., 12:547-550.

34. Randerson, D., 1972, Temporal Changes in Horizontal Diffusion Parameters of a Single Nuclear Debris Cloud, J. Appl. Meteorol., 11:670.

35. Carras, J. N. and D. J. Williams, 1981, The Long-Range Dispersion of a Plume From an Isolated Point Source, Atmos. Environ., 15:2205-17.

36. Gifford, F. A., 1984, The Random Force Theory: Application to Meso- and Large-Scale Atmospheric Diffusion, Proc. Oholo Conference on Boundary Layer Structure, Zichron Ya'acov, Israel, March 25-18, 1984.

37. Larsen, M. F., M. C. Kelley, and K. S. Gage, 1982, Turbulence Spectra in the Upper Troposphere and Lower Stratosphere at Periods Between 2 Hours and 40 Days, J. Appl. Meteorol., 21:1035.

38. Gage, K. S., 1979, Evidence for a $k^{-5/3}$ Law Inertial Range in Mesoscale Two-Dimensional Turbulence, J. Atmos. Sci., 36:1950-1954.

39. Lilly, D. K., 1983, Stratified Turbulence and the Mesoscale Variability of the Atmosphere, J. Atmos. Sci., 40:749-761.

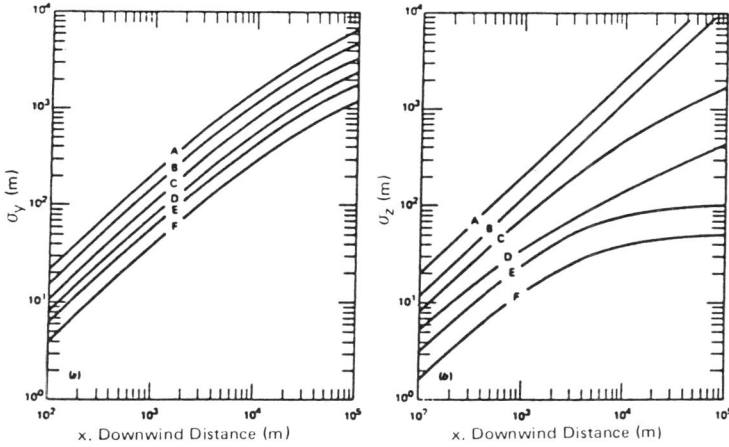

Fig. 1. Curves of σ_y and σ_z, based on recommendations by Briggs for flow over open country (after Gifford[2]).

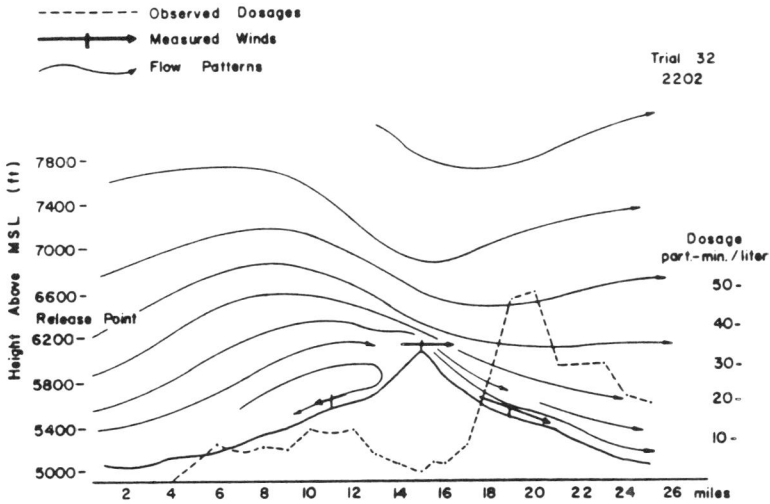

Fig. 2. Schematic flow patterns and surface dosage patterns on a ridge (after Smith[4]).

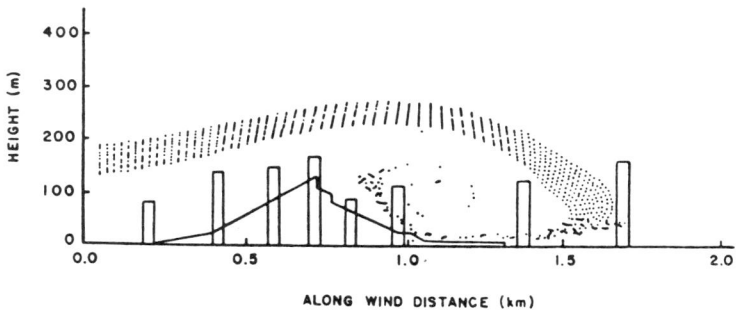

Fig. 3.   Bar graph shows particle "plume" with average observed SO_2 concentrations from a power plant stack.

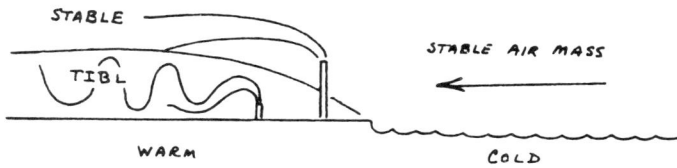

Fig. 4.   Schematic diagram of plume interaction with the thermal internal boundary layer.

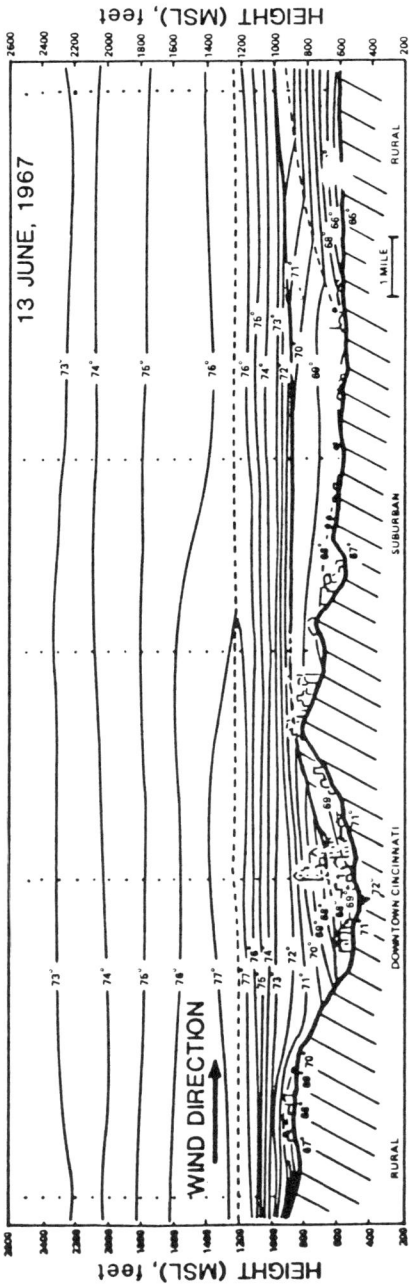

Fig. 5.   Cross section of temperature over Cincinnati, Ohio (taken from Clarke[24]).

CAPABILITIES, NEEDS, AND APPLICATIONS OF GASEOUS TRACERS

Russell N. Dietz and Gunnar I. Senum

Environmental Chemistry Division
Brookhaven National Laboratory
Upton, NY 11973

ABSTRACT

Those compounds which meet most or all of the gaseous conservative tracer criteria for selection, including low atmospheric background concentration, simple sample workup and analysis, high detectability, limited industrial use, harmless, low reactivity and removal rates, and low cost generally are confined to SF_6, halocarbons, perfluorocarbons, and deuterated methanes. The properties, availability, and cost of these tracers are reviewed. A system for tagging and following a parcel of air consists not only of the gaseous tracer, but also its release, sampling, and analysis equipment; each will be reviewed in some detail. For tracers that are gases at normal temperatures, the release is simply through metered valves; for those that are liquids, metered atomization and/or vaporization techniques must be used. Sampling has most commonly been performed by whole air techniques (plastic bags, syringes, evacuated bottles, etc.) but also by adsorbent samplers, located primarily on the ground but in a number of cases above-ground as well with the use of vertical sampling techniques. Analyses for some tracers can be done in real-time making use of aircraft more viable. The techniques for analyzing collected air samples will be discussed. Since the present emphasis today is on long-range (500 to 1,500 km) atmospheric tracing, the needs for enhanced resolution and detectability in the analysis system will be discussed, including their effect on reducing field implementation costs. Application of tracers in atmospheric tracing will predominate the discussion but other applications including indoor air leakage, leak detection, and unique applications (e.g., detection of clandestine bombs) will be mentioned.

I. INTRODUCTION

Gaseous tracers, constituents which can be sensitively, uniquely, and unambiguously detected at very low concentrations in air, have been developed primarily as a result of interest in further understanding and quantifying the nature of atmospheric characteristics such as transport and dispersion. As their capabilities, such as the number of tracers available and the minimum detectable quantity, have increased and improved, not only has

118

the scale and scope of atmospheric transport and dispersion increased (distance of transport measurements now extend to 1,000 km and more) but so has the variety of the uses of these tracers in other applications such as detection of tagged explosives in clandestine bombs, air leakage and mixing within homes and tall buildings for air quality and energy conservation assessment studies, and general leak detection such as condenser tubes in power plants and oil leaks from underground high voltage transmission cables in urban areas.

This paper will provide a brief review of the most prominent types of gaseous tracers available and some typical applications, but will focus primarily on the needs for further research in the particular class of gaseous tracers known as perfluorocarbon tracers (PFTs), which seem to be the most viable in many of the applications. Further, this paper will focus only on gaseous tracers which are conservative, that is, do not react physically or chemically in any way within the environment. Emphasis will also be placed on the application of PFTs to the measurement of transport and dispersion on a 1,000 km scale.

II.   GASEOUS CONSERVATIVE TRACERS

Many types of gases can and have been used to tag and trace the movement of air and other gaseous and liquid fluids from one location to another. Those meeting certain criteria will be discussed in some detail including their properties and costs, how to release the tracers to the air, how to sample the air for the tracers of interest, and finally how to perform the analyses.

A. Criteria for Tracer Selection

For atmospheric studies of transport and dispersion as well as many of the other applications mentioned earlier, gaseous tracers must have the following essential properties:

1. Non-Depositing. During a transport time of up to several days, the tracer must not adsorb onto surfaces nor dissolve in water.

2. Non-Reactive. The tracer must not be affected by photochemical, thermal, aqueous phase, nor any other heterogeneous or homogeneous chemical reactions with a half-life of a few months or less.

3. Low Background. The lower the atmospheric background of a particular tracer, the lower is the amount of tracer that needs to be released in order to have a detectable signal at a given receptor downwind. The atmospheric background may be the tracer's ambient concentration of possibly an equivalent background concentration due to noise or unidentifiable equivalently responding components.

4. Harmless. Neither the users nor the general public must be affected in any adverse way. The tracer should have a low to non-existent toxicity, both from an inhalation and ingestion standpoint, and not exhibit mutagenicity nor carcinogenicity, nor should it cause any adverse environmental impact.

5. Limited Industrial Use. If the source of the tracer being measured is not unique to the release, then the measured concentrations may not unambiguously identify the source-receptor relationships.

6. Available at Low Cost. Unfortunately, the less the industrial use, usually the more expensive is the compound of choice. Thus cost is also a function of the background concentration since the product of the two is related to the actual tracer cost, assuming that the detectability is limited only by the background.

7. Readily and Sensitively Detectable. The tracer should be sampled and analyzed easily and inexpensively, and be detectable down to the current ambient levels at low cost.

Experience has shown that primarily compounds detectable by the electron captive detector (ECD),[1,2] for example, halogenated compounds, and by high resolution isotopic mass spectrometry,[3,4] for example, heavy (deuterated) methanes, make the best candidate gaseous conservative tracers. Each of these materials meet the above criteria to varying extents including ambient background concentrations in the sub-part-per-trillion range. Thus, typical concentration units are pL/L (picoliter per liter), which is equivalent to parts-per-trillion or parts in 10^{12} and fL/L (femtoliter per liter), which is equivalent to parts-per-quadrillion or parts in 10^{15}.

B. Properties and Costs of Tracers

The most widely used intentionally released gaseous conservative tracers meeting essentially all of the criteria listed above are shown in Table 1, listed in order of decreasing ambient concentration. All of those are detectable by the ECD with the exception of the last two which are measured by mass spectrometry.

Sulfur hexafluoride (SF_6) was one of the first ECD-sensitive compounds to be used because its response was very high (one electron reacted with each and every molecule), it was available as a liquified gas in large cylinders at reasonable cost ($\sim$ $10/kg), and its low boiling point (and hence its high vapor pressure) simplified the controlled release from the cylinders. The most unique property, however, was the ability to elute SF_6 ahead of all other atmospheric constituents during its gas chromatographic (GC) determination on a 5A molecular sieve column.[5]

The primary use of SF_5 is as an electrical insulating gas for high voltage power switching equipment at the rate of about 2,000 tons/year, from which it is ultimately released to the atmosphere where it permanently resides. As shown in Table 2, local concentrations of SF_6 can be as high as 2,000 fL/L, whereas the tropospheric background is only 850 fL/L.

Halocarbons 13B1 ($CBrF_3$) and 12B2 (CBr_2F_2) have also been used as tracers, with the 13B1 having physical properties very similar to SF_6 (cf. Table 1); but its ECD sensitivity is 20- to 50-fold lower than that of SF_6. Both SF_6 and F13B1 can be separated on the same 5A molecular sieve GC column.[5] The detection of ambient F12B2 has been limited by the analytical resolution, but it is probably less than 20 fL/L.

The four perfluorocarbon compounds shown in Table 1, PDCH, PMCH, PMCP, and PDCB are liquids at room temperature and thus their release is more difficult, requiring atomization or vaporization. They have essentially the same ECD sensitivity as SF_6, but, as shown in Table 2, significantly lower background concentrations.

Based on typical meteorological conditions and experience from past tracer experiments, the quantity of tracer that must be released for a 3-h period in order to detect a peak centerline concentration of 100 times background at 100 km away is listed in Table 2 for each tracer. When the quantity released is multiplied by the tracer cost per unit weight, the relative tracer cost for the experiment is determined. The last column in Table 2 shows that SF_6, F13B1, and PDCH all have about the same relative cost. The F12B2 appears to be an order-of-magnitude lower in cost, but a simple sampling and analysis scheme has not been developed.

The next three PFTs appear to have the lowest relative tracer costs with the exception of the heavy methanes. PMCH and PMCP are both commercially available from I.S.C. Chemical Limited, England, but the PDCB is no longer commercially available. The previous manufacturer, DuPont, is, at present, not interested in making this compound and the other companies such as 3M in the U.S. and ISC Ltd in England cannot manufacture this or similar cyclobutane compounds. As will be shown later, because all four PFTs can be sampled from the air and analyzed by a single tracer system, they form a powerful tracer class.

The last two tracers, the heavy methanes, although very expensive per unit weight require very little tracer to provide a signal above the detection limit. Thus, the tracer cost is the lowest of all those listed. However, the cost of the analyses and the time required per sample is much greater for the heavy methanes than the PFTs. The heavy methanes are more suitable for continental and global scale experiments when measurements must be made at extremely low concentrations, and relatively small number of samples may provide the desired information. When the scale of transport is from 100 to 2,000 km and/or a large number (several thousand) of samples are required, the PFT system is preferred.

C. Tracer Release

Details on how to estimate the quantity of tracer that needs to be released based on both meteorological models and previous experience plus a description of the release equipment for both the gaseous and liquid tracers have been described elsewhere.[5] Basically, the gaseous tracers are released from liquified gas or compressed gas cylinders, which are weighed before and after (or during) the release, and metered through a gas flow measuring device. The liquid tracers are dispersed either with a high pressure metering pump and vaporized into a stream of hot air from an auxiliary blower or are atomized directly into the air with a paint-type compressor spraying equipment.

III.   SAMPLING, ANALYSIS, AND APPLICATIONS OF GASEOUS TRACERS

Since the sampling, analysis, and applications of the various tracers is at many times a system approach, they will be discussed in this section in that manner. Examples of the use of the heavy methanes, SF_6, and perfluorocarbons will be given.

A. Heavy Methanes

The heavy methane system was developed and implemented by Los Alamos National Laboratory.[3]

1. Sampling. Plastic bag sampling is by far the most common means for sampling all types of gaseous tracers from collected whole air samples. This includes the heavy methanes, SF_6, halocarbons, and, to a lesser extent, the PFTs. Figure 1 shows a typical programmable bag sampler being readied for a heavy methane field tracer experiment. In this case, only 1 large bag capable of collecting up to 30 to 50 L of ambient air, is being used because of the need to collect sufficient quantity of the methane tracer. Alternatively, cryogenic samplers extract methane (both normal methane present at 1.5 ppm as well as the tracer methane) from the air by adsorption on activated charcoal maintained at liquid nitrogen temperature.[6] Bag samples are transferred to metal cylinders containing an added 1 ml of normal CH_4 prior to shipment back to the laboratory for analysis.[4,5]

2. Analysis. Following sample collection in bags and transfer to CH_4-spiked metal cylinders, the sample is analyzed for normal methane concentration; for the cryogenic samples without added methane, the normal ambient concentration of 1.5 ppm is assumed. Then the methane fraction is separated and purified by a preparative gas chromatograph system (cf. Fig. 2) and analyzed by the mass spectrometer (cf. Fig. 3). The time to perform a single analysis is thus about an order of magnitude longer, perhaps 1 to 2 hours, than the analysis of the ECD sensitive tracers.

3. Application. In a test of the two heavy methanes with SF_6 and krypton-85 co-released for 4 h at the Savannah River Laboratory, approximately 30-L air samples were collected over a 7 h period at 100 km downwind.[4] As shown in Fig. 4, the relative shapes of each of the crosswind concentration profiles were nearly identical. The ratio of the methanes to SF_6 found was only 25 to 35% of that in the release ratio but the SF_6 to krypton-85 agreed with that released; the cause of the discrepancy was never entirely determined.[7]

B. Sulfur Hexafluoride and Halocarbons

1. Sampling. SF_6 is essentially always sampled by a whole air sampling technique such as the automated plastic bags discussed above or by a number of other techniques including syringes (manual and automated), plastic squeeze bottles, and evacuated bottles (manual and automated).[5] Field adsorption sampling onto 0.1 g of charcoal was also implemented for SF_6, but whole air sampling is the most practical approach. Halocarbons are sampled by the same whole air techniques.

2. Analysis. These tracers are generally analyzed by small portable gas chromatographs (GCs) brought into the field for rapidly analyzing the numerous whole air samples brought back to the field laboratory. In some cases, the samples are shipped back to a central laboratory and analyzer. A number of portable GCs are commercially available (e.g., Systems, Science, and Software, LaJolla, CA; ITI, Inc., Burlington, MA) for rapid analysis of whole air collected SF_6 samples at 5 pL/L and above. Greater precision and a lower detection has been achieved by the use of pre-cut column technique[8] and a 13X molecular sieve concentrator;[9] with these techniques, ambient concentrations at 0.5 pL/L were measured from 40 to 80 mL air samples with a precision of $\pm$ 2%.

Ambient halocarbon measurements have been made by a number of researchers, many of whose techniques would be applicable to the halocarbon tracer[10,11]

3. Application. The literature is replete with examples of the use of SF_6 to measure the transport and dispersion of air pollutants at distances up to about 100 km. As shown in Fig. 4, SF_6 was measured from bag samples by the pre-cut column and concentration techniques over the range from ambient (~0.5 pL/L) up to a peak value of almost 6 pL/L during the Savannah River Plant experiment. The standard deviation of the width of the plume (σ_y) was estimated to be 7.4 km which was equivalent to a class C ($\sigma_o = 15°$) at 100 km downwind on an extrapolated Pasquill-Gifford plot.

SF_6 was released at the rate of 39 kg/h; it is quite apparent that in order to measure a peak concentration of 100 times background, the release should have been about 5.7-fold higher (correcting for the 4 h release and 7 h measurement period), that is, equivalent to 220 kg/h, essentially identical to the hourly rate quoted in Table 2 of 193 kg/h when correcting for the 4-fold change in the ambient SF_6 concentration. Note that the current

tropospheric, SF_6 background is increasing at the rate of almost

0.1 pL/L/year.

C. Perfluorocarbons

1. Sampling. The PFTs can be collected and analyzed from whole air

samples as are SF_6 and heavy methanes, but two types of adsorbent samplers

have been specifically developed for the PFTs, a programmable sampler[12]

and a passive sampler.[13]

a. Programmable PFT Sampler. This was developed by Dietz at

Brookhaven and commercially manufactured for the National Oceanic and

Atmospheric Administration (NOAA) by Gilian Instrument Corporation as the

Brookhaven Atmospheric Tracer Sampler (BATS). The entire unit measures

just 14 x 10 x 8 inches and weighs 7 kg (cf. Fig. 5). The lid contains 23

sampling tubes, each containing 150 mg of 20-50 mesh type 347 Ambersorb

(Rohm and Haas Co.), which can retain all the PFTs in more than 30 L of air.

Internal batteries provide power for up to one month of unattended operation

of all the automatic sampling and recording features. A summary of features

is given in Table 3.

Sample recovery was accomplished by direct ohmic heating of the

adsorption tube to 400°C, with the PFTs being purged from the BATS tube

through an automated ECD-GC system as described in Part 2 following. All

23 tubes were automatically analyzed in about 3 h.

b. Passive PFT Sampler. Originally developed as a means to

measure the indoor PFT concentration during the determination of air

infiltration and air exchange rates in homes and buildings using miniature

PFT permeation sources,[13] the passive sampler has also been used in

atmospheric tracer studies.

In its first configuration, one end of the sampler contained a 1 mm

capillary tube and so was coined the Capillary Adsorption Tube Sampler

(CATS). The present configuration of the passive sampler, which is made

from 6 mm OD by 4 mm ID glass tubing exactly 2.5 inches (6.4 cm) long,

contains 64 mg of Ambersorb 347. Sampling occurs by the process of Fickian

diffusion when one cap is removed. From the depth to the bed (2.76 cm), the

cross-sectional area (0.126 cm^2), and the empirically derived diffusion

coefficients of the PFTs in air, it was determined that the CATS sampled at

a rate equivalent to about 200 mL of air per day for PMCH. In some recent comparisons between BATS and CATS, the actual sampling rate was found to be equivalent to 232 mL air/day for PMCH and 217 mL air/day for PDCH. With two caps removed, increasing wind speed causes increasing errors. The percent detection limit is about 0.1 part-per-trillion-hour (i.e., 100 fL-h/L), adequate to detect ambient PMCH (3.6 fL/L) in a one-month integrated sample. Table 3 summarizes the CATS features.

Ambient concentration measurements of PMCH and PDCH were recently determined with 10 CATS deployed in the Washington, D.C. area for a period of 30 days. As shown in Table 4, the background concentration of both tracers was comparable to that shown in Table 2.

2. Analysis of PFTs. Whether from programmable or passive samplers, the sample is automatically thermally desorbed and passed through a Pd catalyst bed, permeation dryer, and a pre-cut column before being reconcentrated on an in situ trap. The trap prevents the collection of unwanted low molecular weight constituents, the pre-cut column prevents the passage of unwanted high molecular weight constituents, the catalyst bed in the presence of 5% H_2 in the N_2 carrier gas destroys interfering freons and oxygen, and the dryer removes moisture from the ambient samples. After thermally desorbing the trap, it is injected into the main column after passing through another Pd catalyst bed and finally the ECD. Limits of detection range from 0.5 to 5 fL up to maximum of about 5,000 pL or, roughly 6 orders of magnitude.

Initially the chromatographic column was 20 feet of Porosil F, and an example chromatograph of the three earliest used tracers, PDCB, PMCH, and PDCH separated on Porosil F is shown in Fig. 6. A is apparent in this figure, three is nearly the maximum number of tracers that can be simultaneously analyzed with the resolution available with this column. Recently, a new column packing has been introduced, i.e., Carbopack C with 0.1% SP-1000, which is the presently used packing. Figs. 7 and 8 show the increase in resolution for these three tracers, including separation of the three isomers (meta, para, and ortho) of PDCH, and a newly identified tracer, PMCP. The PDCB at a column temperature of 140°C was eluted just ahead of the PMCP peak, which was labelled as the second unknown in Figs. 7 and 8; at 100°C, the PDCB was clearly separated from the PMCP. Table 5 gives the results of 6 background air samples analyzed at the 140°C column temperature; the reproducibility was within ±10% even for the sub-fL/L PDCB concentration.

Since ambient air samples currently contain uniform tropospheric background concentrations of many PFTs, once resolved and quantified, any one of them can become an internal refe·ence standard to determine the volume of air sampled, thereby providing redundancy in the recording of the sampled volume size and a means of quality assurance for sampler reliability.

3. Applications. The PFT programmable sampler/laboratory GC system was tested in a 600 km tracer experiment in which tracer was released from Norman, Oklahoma, in July 1980 and samples collected on the ground with 20 BATS along a 100 km arc and 40 BATS along a 600 km arc as shown in fig. 9. Before deployment, the BATS must be thoroughly checked out for performance (cf. Fig. 10) and each tube baked out for 10 min. The results obtained for 3-hour integrated samples collected at the 600 km sites as shown in Fig. 11, gave peak concentrations nearly 3 order-of-magnitude above ambient levels for PMCH which was released at the rate of 64 kg/h.[12] The plume σ_y was about 29 km, which, at 600 km downwind, was also equivalent to a Class C dispersion.

In addition to their use as atmospheric tracers, the PFTs have been used successfully in tagging electric blasting caps during their manufacture such that, if used in clandestine bombs, the presence of the exploding device could be detected with a continuous tracer analyzer.[14] Another area that is growing at a moderately significant rate is the use of PFTs in the detection of air leakage or infiltration rates into homes and buildings during the heating season. passive PFT sources and passive samplers of the type described previously have now been employed in more than 1,000 homes.[13] Other leak detection applications which are underway include the detection of oil leaks from underground high voltage oil-filled transmission cables in high density urban areas. A small amount of PFT (<0.1 wt. %) dissolved in the oil will be detectable above ground with sampling and real-time PFT analyzers to help locate the leaks.

IV.   SPECIAL SAMPLING AND REAL-TIME ANALYSIS EQUIPMENT

In addition to ground sampling requirement during transport and dispersion experiments with gaseous tracers, it is also essential to perform measurements of the tracer concentration aloft in order to determine such things as the height of the mixed layer and the shear of the plume with altitu.le. Such information is required if a complete tracer material balance is to be performed and if current atmospheric transport and dispersion models are to be validated and improved. Special systems which have been developed and field tested to perform these measurements include a real-time continuous tracer analyzer, a real-time dual-trap PFT analyzer, and several vertical atmospheric sampling cables.

A. Real-Time Continuous Tracer Analyzer (CTA)

The CTA in its present configuration was conceived by J. Lovelock,[15] in which air was continuously mixed with half as much hydrogen to catalytically, over a Pd-on-molecular sieve catalyst bed, convert all of the oxygen to water, which was removed by a condenser, and to destroy unwanted interferents. The tracers, such as SF_6 or PFTs, survived the catalyst bed and in the remaining N_2 of the original air sample were continuously measured in an ECD. Subsequently, numerous improvements have been made including the use of a permeation-type Nafion dryer,[16] mass flow controllers to reduce altitude effects[17,18] and high-throughput pumps to reduce delay and response time.[18]

These continuous analyzers have been used in a number of short (about 10 km) range[17,19,20] and moderate (up to 100 km) range SF_6 dispersion tests. Two such prototype units were flown on a small plane (cf. Fig. 12) which made passes at various altitudes at 12, 31, and 78 km downwind of the release point at Versailles, IN, on October 7, 1977 (cf. Fig. 13). The results from several of the passes are shown in Fig. 14 and the computed standard deviations of the width of the plume were compared with the Pasquill-Gifford curves as shown in Fig. 15.

The CTA is an extremely powerful tracer analyzer for measuring crosswind concentration profiles in the range of peak concentrations from 100 to 50,000 pL/L. Because the limit-of-detection for SF_6 and PFTs is about 10 pL/L, the instrument is not useful for long range (greater than 500 km) PFT dispersion experiments, where peak concentrations are primarily in the sub-pL/L range.

B. Dual-Trap Real-Time PFT Analyzer

The peak PFT concentration typically encountered in long range tracer experiments range from 1 to 5 pL/L at 100 km downwind to about 100 to 400 fL/L at 800 km downwind. These concentrations can easily be measured with BATS located both on the ground and in aircraft. The expense of an aircraft platform, however, mandates that the sampling occur where the plume is actually located. Thus, the need exists for a real-time PFT analyzer.

Originally conceived and developed by Lovelock, an instrument was modified by Brookhaven to concentrate the PFTs in a 5-L air sample and analyze the desorbed sample on an in situ chromatograph column and detector. This version had two ambersorb traps. While one was sampling at 1 L/min for 5 min, the other was thermally desorbed and analyzed as shown in Fig. 16. Since the traps reversed position every 5 min, no tracer was lost. The limit of detection was about 10 fL/L making it quite useful for potential aircraft traverses in long range (1,000 km) PFT experiments. At the very least the dual-trap analyzer would direct the aircraft for on-board programmable sampler collection of the otherwise invisible plume.

A new version of the dual-trap analyzer was recently completed for use in the 1983 CAPTEX tests in September and October.[21,22] In these series of 7 separate tests, PMCH was released from Dayton, OH (5 times) and Sudbury, Ontario (2 times). Nearly 80 ground sampling stations equipped with BATS throughout the northeastern U.S. and southeastern Canada region were used to measure the transport of the plume on the ground and up to seven aircraft equipped with modified high flow rate (500 mL/min) BATS made crosswind traverses of the plume at different distances downwind (from 300 to 825 km) and at several altitudes.

On-board one aircraft was the new dual-trap analyzer. This unit revolves 3 tracers in a 4-min chromatogram of a 4-min air sample collected at the rate of 1 L/min. For CAPTEX, the conditions were set to monitor the concentrations of ambient PMCH and PMCP, which are about 3.6 and 2.7 fL/L, respectively. Ultimately, one tracer will be a reference tracer, since it will not be released during the experiment. An internal microprocessor will automatically compute the response of each released tracer relative to the referenced ambient tracer in order to display the concentration of each directly on an analog output. Eventually, a commercial manufacturer for this analyzer should be sought.

A map of the CAPTEX region is shown for Mission 5 in Fig. 17; pass 1 from the 3rd sortie is shown by the line, and the arrow represents the centerline of the plume of tracer released from Sudbury (~200 kg of PMCH in 3 h) as found by the real-time analyzer on-board the aircraft. The arrival of the plume between 1,500 and 1,800 hours is shown by the passes in Fig. 18, with full development of the plume shown in the 3 passes of Fig. 19. The crosswind standard deviations for 4 of the 6 aircraft missions were computed and plotted in Fig. 15; the data fell in a reasonable area of such a plot.

Similar information will be obtained from integrated tracer data obtained with the programmable sampler, BATS, taking sequential 10-min samples as the plane traversed the plume, although from Figs. 18 and 19 it is apparent that only about seven 4-min integrated values are obtained in a typical traverse. Ten-min BATS samples would and did only provide 2 or 3 values for each traverse. Thus, for airborne traverses even at 1,000 km downwind, 3 or 4-min analyses are necessary.

C. Vertical Atmospheric Sampling Cables (VASC)

Although aircraft platforms with real-time analyzers can provide useful pictures of the plume aloft, in certain complex terrain tracer studies, for example, near mountains and valleys and especially at nighttime, the well-defined directions of the valley flows lend themselves to be monitored with balloon-borne vertical sampling devices for the vertical definition of the transport and dispersion of pollutants.

Several types of balloon-borne systems have been designed and used in the ASCOT program (Atmospheric Studies in Complex Terrain) in which tracers, including SF_6, heavy methanes and perfluorocarbons, were measured in air samples collected to a height of 500 m above the ground.[23,24] Traditionally, these systems have used tethered balloons, from 5 to 100 m^3 in volume to suspend whole air sampling packages such as plastic bags with pumps and automated syringes at several altitudes along the cable or to collect sequential whole air samples over time and space as the balloon is moved from the ground to its maximum altitude.

In this section, a brief description will be given of two novel balloon-borne VASCs - a multitube, ground-based sampling pump, cable (VASC-I) and a single-tube, multi-level absorbent sampler, cable (VASC-II) with a single pump on the ground.

1. VASC-I. Four separate 1/4-inch OD polyethylene sampling cables were bundled together in a braided Kevlar sheath (Cortland Line Company) and suspended from a $100-m^3$ balloon tethered for an altitude of 1,600 ft above the ground.[24] The cable was designed at Brookhaven to sample four separate 400-ft layers from 0-400 ft up to 1,200-1,600 ft as shown in Fig. 20. Each cable had intake holes at 50-ft intervals over the designated sampling span. Pumps on the ground pulled the air through each cable to the programmable samplers, bag samplers, and a real-time dual-trap analyzer.

About 445 g of PDCH were released about 4 km upwind of the tethered balloon site which was at a 1,250-ft lower elevation in the complex mountainous terrain of the Geysers area north of San Francisco, CA. The release was made for 1 hr commencing at midnight in order to study the drainage flows from the mountains into the valleys. The vertical cable allowed the concentration distribution aloft to be recorded on an hourly basis with the BATS programmable sampler and the dual-trap analyzer as shown in Fig. 21.

2. VASC-II. A second vertical sampling cable is currently being prototyped. It consists of a 500-m long polypropylene tubing, 0.062-in. ID by 0.010 in. wall, with 15 orifices located 50-m apart from 500 m down to 200 m and then decreasing to only 10 m apart near the ground. The orifices have been individually sized according to a model to allow the air to be sampled at closer intervals near the ground, but at the same flow rate, about 4.5 ml/min, through each sampling tube at every altitude. Each orifice will be equipped with a passive sampler of the type described in Section 3.C.1.b, but used in an active mode. Flow through each sampling tube will be provided with a single small pump, pulling a total of 67.5 ml/min at about 0.5 atm vacuum, located on the ground. Thus a 2-h sample will represent slightly more than 0.5 L of air in each tube, sufficient to measure the background of PMCP and PMCH with about $\pm 10\%$ precision.

Fig. 22 shows a schematic representation of the cable, which will be flown in the September-October 1984 ASCOT experiment in Colorado for the first time, collecting vertical time-integrated samples for 2-h periods about every 3 h. The entire cable with sampling tubes weighs less than 1.4 pounds (0.64 kg) and is capable of flying to 500 m above ground level with a 3 m^3 balloon. In Colorado, the Brookhaven site will have a ground elevation of

6,250 ft (1.90 km), and thus a 5 m^3 balloon will be used to provide the necessary lift for the altitude transmitter as well.

V.   PRESENT CAPABILITIES AND FUTURE NEEDS

Gaseous tracer technology has advanced markedly in the last decade in terms of both new and unique sampling and analysis tools as well as new tracer gases, such as the heavy methanes and perfluorocarbon tracers (PFTs), especially in their application to studies in complex terrain and in long range transport and dispersion where many thousands of samples are collected and analyzed. It is this latter requirement which, for many of these applications, precludes the use of the heavy methanes; they are ideal when only a limited number of analyses can provide the necessary information such as in continental–scale and global transport.[25]

Since the PFTs appear to be tracers of the future, this section will provide a summary of the present capabilities and future needs in the development of PFTs as a viable tracer system including research into additional PFTs and improvements needed in the programmable sampler, the laboratory gas chromatograph system, and the real–time dual–trap analyzer.

A. Available PFTs

As indicated earlier, the family of four PFTs shown in Tables 2 and 5 represent a potentially powerful system for experiments in which multiple release sites are to be tagged and unambiguously identified at downwind receptor sites. One area in which this tracer system will play an invaluable role is in the presently contemplated year–long massive aerometric tracer experiment (MATEX).[26] The two PFTs that can be used cost–effectively in such an experiment are listed in Table 6 along with research steps that need to be taken to increase the number of available tracers to the desired 6 to 8.

With the analytical precision of the present GC system and the background concentration of the current 2 tracers, a minimum tracer release rate of 50 kg/h is required for achieving detectable (5 times ambient) concentrations at 1,000 to 1,400 km downwind. The tracer specifications and costs for such a year–long experiment are given in Table 7. The projected total cost of such an experiment is about $120 million and thus the tracer cost is a significant ($15 million) portion. Research must be undertaken to develop additional PFTs with costs less than $100/kg that have background concentrations less than 1 fL/L, and that can be sampled and analyzed on the same PFT equipment.

B. The Programmable Sampler (BATS)

The programmable sampler shown in Fig. 5 has allowed a number of long-range and complex terrain tracer experiments to be conducted in a cost effective manner. But as a result of their repeated use, experience has been gained into the problems frequently encountered and the features that, if present, would greatly facilitate future experiments and reduce time and manpower costs while improving the percentage of useful data.

Table 8 shows that the problems have centered around the pump and data recording features. Occasionally units would develop electronic failures and operation would sometimes have difficulty following the programming sequences. New units should have improved pumps, remote programming and performance testing, and more flexibility in programming and data handling.

C. The Laboratory Gas Chromatograph (GC) System

The programmable sampler can be analyzed on the present GC system as shown in Fig. 23 with the specifications given in Table 9. However, as was stated earlier, the present packed column technology with isothermal column temperature will only allow 3 or 4 tracer peaks to be resolved and the analysis time is about 10 min per chromatogram or 4 h per lid assembly (23 tubes).

In order to handle the many analyses that would be required in a large scale field experiment (e.g., over 500,000 samples would need to be analyzed in a MATEX-type experiment[26]), research needs to be applied in those areas indicated in Table 9. If high resolution capillary column or microparticulate column GC is implemented, it has been shown that the amount of tracer that needs to be released in order to quantitatively measure its contribution 1,400 km away, can be reduced by 20- to 50-fold, thus, not only saving more than $10 million, but also bringing the quantity of tracer needed within the current capacity of commercial suppliers.[26] To provide the ultimate in GC technology as a result of the exploration of each of the items listed in Table 9 would require a minimum of 1.5 to 2 man-years of research effort.

D. The Real-Time Dual-Trap Analyzer

The capabilities of the present real-time PFT analyzer are summarized in Table 10. Using technology developed for more rapid analyses in the laboratory GC system, adaptations need to be made to provide more rapid thermal recovery and shorter elution times for multiple PFTs (4 or more). A quality commercial supplier of this instrument will be needed.

ACKNOWLEDGMENT

Special appreciation is expressed to a number of colleagues who have contributed over the course of years. Ed Cote has continually dedicated himself to the implementation of the laboratory GC techniques for PFTs. Bob Goodrich has played a valuable role in the development of the continuous tracer analyzer and computerizing the analysis data handling. Bob Wieser has helped with the production of passive samplers and vertical sampling cables. Ted D'Ottavio fabricated and employed the real-time dual trap PFT analyzer.

REFERENCES

1. J. E. Lovelock and M. L. Gregory, Gas Chromatography, p. 219, Academic Press, New York, 1962.

2. C. A. Clemons and A. P. Altschuller, Responses of electron-capture detector to halogenated substances, Anal. Chem: 38 (1), 133-136 (1966).

3. G. A. Cowan, D. G. Ott, A. Turkevich, L. Machta, G. J. Ferber, and N. R. Daly, Heavy methanes as atmospheric tracers, Science 191, 1048-1050 (1976).

4. M. M. Fowler, The use of heavy methane as long range atmospheric tracers, LA UR-80-1342, Los Alamos National Laboratory, Los Alamos, NM, September 1979.

5. R. N. Dietz and W. F. Dabberdt, Gaseous tracer technology and applications, BNL, 33585, Brookhaven National Laboratory, Upton, NY, July 1983.

6. M. M. Fowler and S. Baur, A long range atmospheric tracer field test, Atmos. Environ. 17 (9), 1677-1685 (1983).

7. T. V. Crawford, Ed. Heavy methane-SF$_6$ tracer test conducted at the Savannah River Plant, December 10, 1975, DP-1469, Savannah River Laboratory, Aiken, SC, April 1978.

8. R. N. Dietz and E. A. Cote, GC determination of sulfur hexafluoride for tracing air pollutants, Amer. Chem. Soc., Div Water, Air, and Waste Chemistry 11, 208–215 (1971).

9. R. N. Dietz, E. A. Cote and R. W. Goodrich, Air mass movements by real-time frontal chromatography of sulfur hexafluoride. In Measurement, Detection and Control of Environmental Pollutants, IAEA, Vienna, Austria, September 1976.

10. H. B. Singh, L. J. Silas, and R. E. Stiles, Selected man-made halogenated compounds in the air and oceanic environment, J. Geophhys. Res. 88 (C6), 3675–3683 (1983).

11. D. Lillian, H. B. Singh, A. Appleby, and L. A. Lobban, Gas chromatographic methods for ambient halocarbon measurements, J. Environ. Sci. Health A11, 687–710 (1976).

12. G. J. Ferber, K. Telegadas, J. L. Heffter, C. R. Dickson, R. N. Dietz, and P. W. Krey, Demonstration of a long-range atmospheric tracer system using perfluorocarbons, ERL ARL-101, National Oceanic and Atmospheric Administration, Rockville, MD, April 1981.

13. R. N. Dietz and E. A. Cote, Air infiltration measurements in a home using a convenient perfluorocarbon tracer technique, Environ. Internl. 8, 419–433 (1982).

14. G. I. Senum, R. P. Gergley, E. M. Ferreri, M. W. Greene, and R. N. Dietz, Final report of the evaluation of vapor taggants and substrates for the tagging of blasting caps, BNL 51232, Brookhaven National Laboratory, Upton, NY, March 1980.

15. P. G. Simmonds, A. J. Lovelock, and J. E. Lovelock, Continuous and ultrasensitive apparatus for the measurement of airborne tracer substances, J. Chromatrogr. 126, 3–9 (1976).

16. R. N. Dietz and R. W. Goodrich, The continuously operating perfluorocarbon sniffer (COPS) for detection of clandestine tagged explosives, BNL 28114, Brookhaven National Laboratory, Upton, NY, May 1980.

17. R. A. Baxter, D. Pankratz, and I. Tombach, An advanced continuous SF_6 analyzer for real-time tracer gas dispersion measurements from a moving platform, AV-TP-83/504, Aerovironment, Inc., Pasadena, CA, May 1983.

18. R. Benner and B. Lamb, An improved continuous SF_6 analyzer, Washington State University, Pullman, WA, May 1984.

19. W. F. Dabberdt, R. Brodzinski, B. C. Cantrell, R. E. Ruff, R. N. Dietz, and S. Sethu Raman, Atmospheric dispersion over water and in the shoreline transition zone, Project 3450, SRI International, Menlo Park, CA, December 1982.

20. W. F. Dabberdt, et al., central california coastal air quality model validation study: data analysis and model evaluation, Project 3868, SRI Inernational, Menlo Park, Ca, February 1983.

21. T. D'Ottavio, R. W. Goodrich, and R. N. Dietz, Perfluorocarbon measurement using an automated dual trap analyzer, report in press, Brookhaven National Laboratory, Upton, NY, August 1984.

22. G. S. Raynor, R. N. Dietz, and T. W. D'Ottavio, Aircraft meaurements of tracer gas during the 1983 cross Appalacian tracer experiment (CAPTEX), BNL 35035, Brookhaven National Laboratory, Upton, NY, July 1984.

23. R. Woods, SNL vertical profiling of tracer concentrations, in ASCOT Data from the 1980 Field Measurement program in the Anderson Creek Valley, CA, UCID-18874-80, Lawrence Livermore National Laboratory, Livermore, CA, April 1983, pp. 1468-1475.

24. G. J. Ferber, K. Telegadas, C. R. Dickson, P. W. Kery, R. Lagomarsino, and R. N. Dietz, ASCOT 1980 perfluorocarbon tracer experiments, in Ibid., pp. 1202-1316.

25. E. J. Mroz, Tracking antarctic winds, LA-UR-83-1419, Los Alamos National Laboratory, Los Alamos, NM, 1983.

26. R. N. Dietz and G. I. Senum, Feasibility of perfluorocarbon tracers (PFTs) in source-receptor experiments, unpublished position paper, Brookhaven National Laboratory, Upton, NY, March 1984.

Figure 1.   A programmable sequential bag sampler.

Figure 2.   Preparative gas chromatograph system at LANL for separation of methane from air constituents.

Figure 3.   Mass spectrometer at LANL for determination of heavy methane to CH_4 ratios.

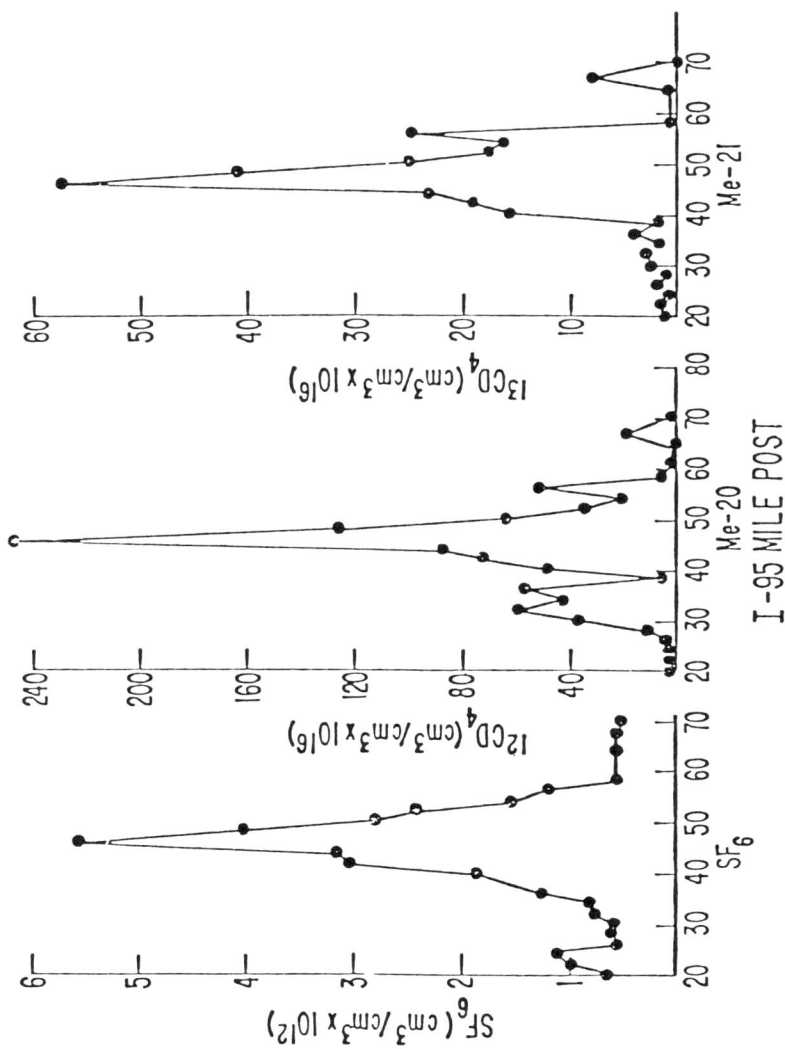

Figure 4.   Bag sample 7-hour integrated ground level SF$_6$ and heavy methane concentrations at 100 km downwind.

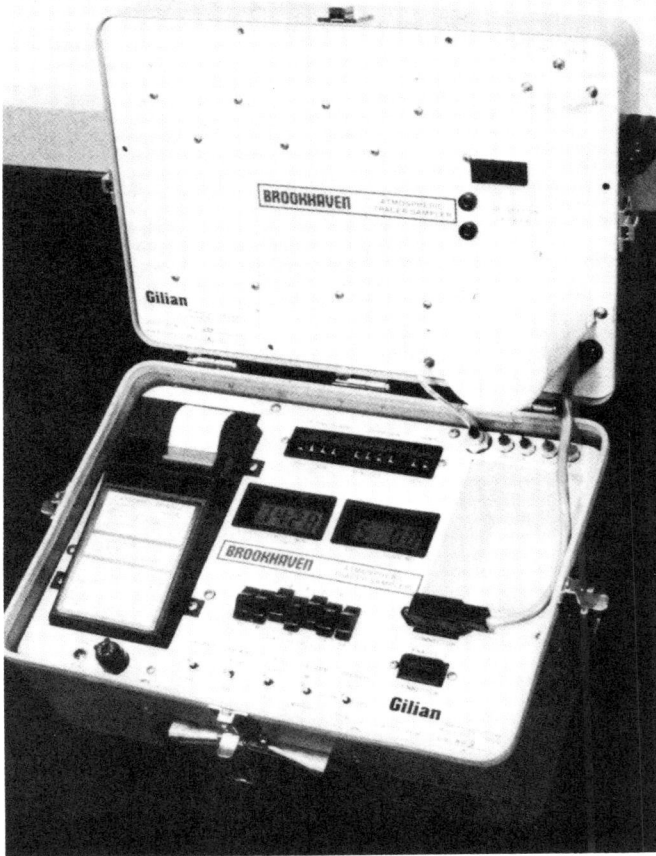

Figure 5.   Brookhaven atmospheric tracer sampler (BATS).

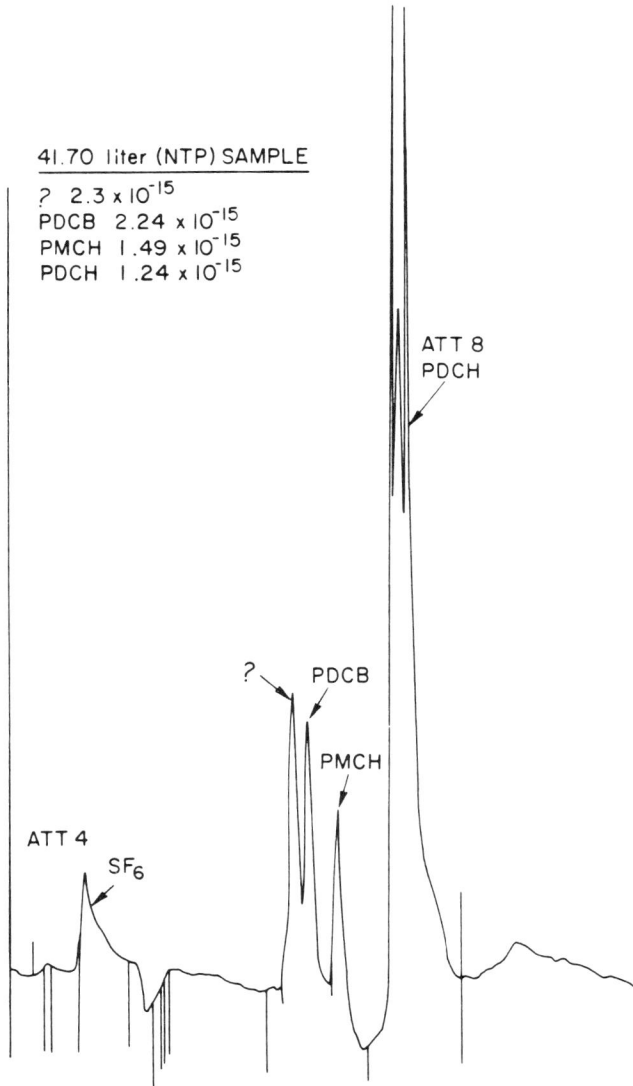

Figure 6.   Chromatogram of 24-L Long Island air sample on a 20-ft porasil F column at 85°C with carrier gas at 10 mL/min.

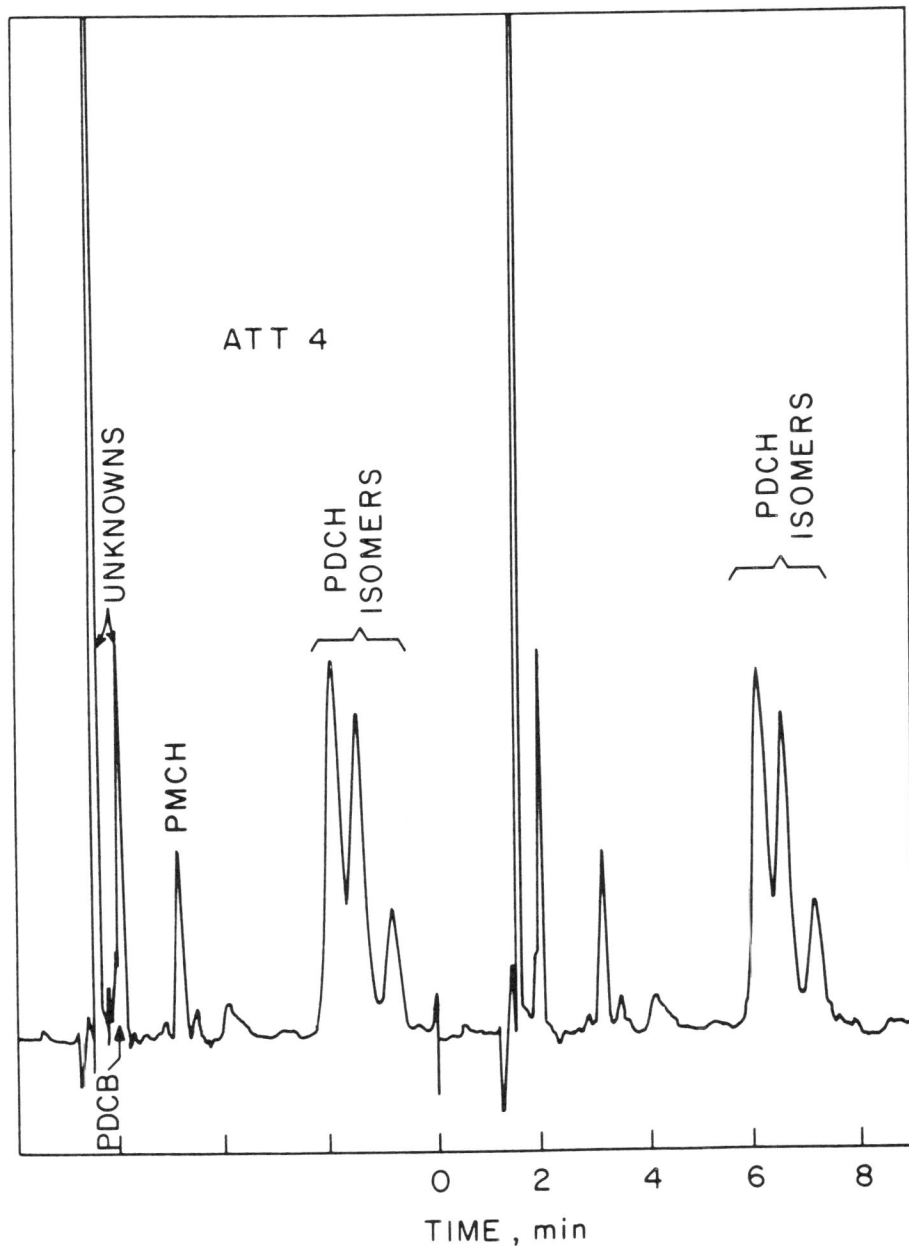

Figure 7.   Chromatogram of 25-L Long Island air samples on a 4-ft column of carbopack C with 0.1% SP-1000 at 140°C with 5% H_2 in N_2 carrier gas at 22 mL/min.

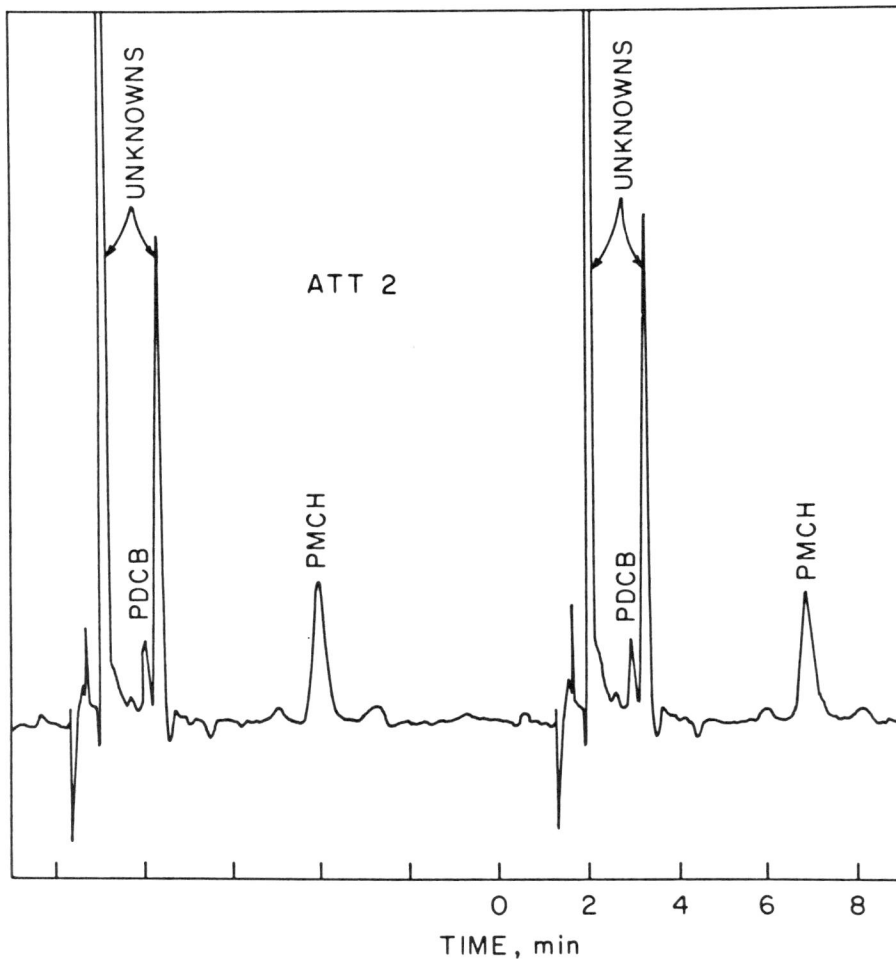

Figure 8.   Same as Figure 7 but with column at 100°C.

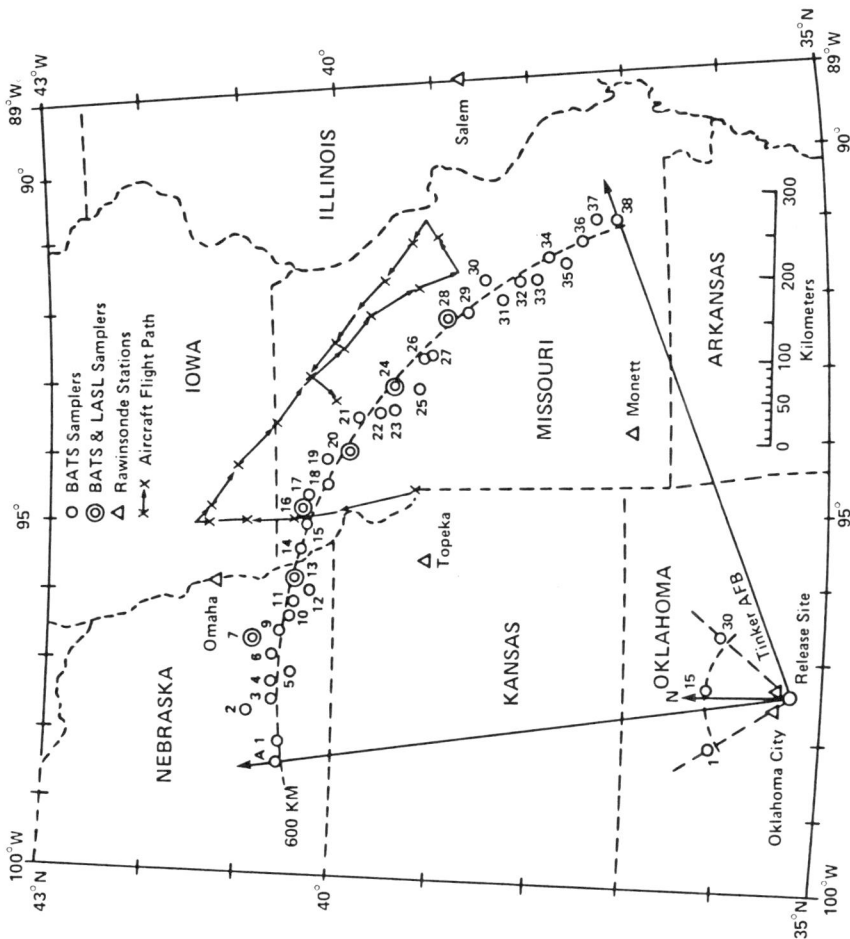

Figure 9.   Location of PFT samplers (BATS), LANL heavy methane samplers and aircraft sampling flight path at 600-km arc.

Figure 10.   BATS perfluorocarbon tracer samplers being checked for performance prior to field deployment.

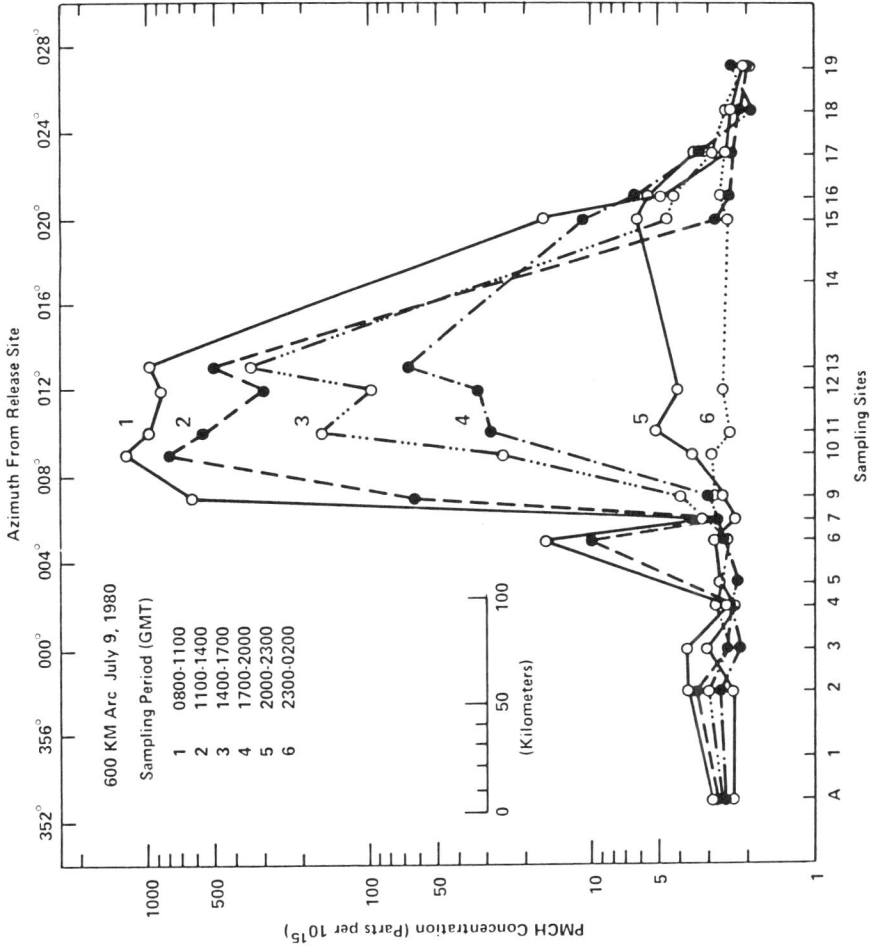

Figure 11.   Integrated 3-hour PMCH concentrations along the 600-km arc.

Figure 12.   Prototype SF_6 continuous tracer analyzers aboard a small Cessna aircraft.

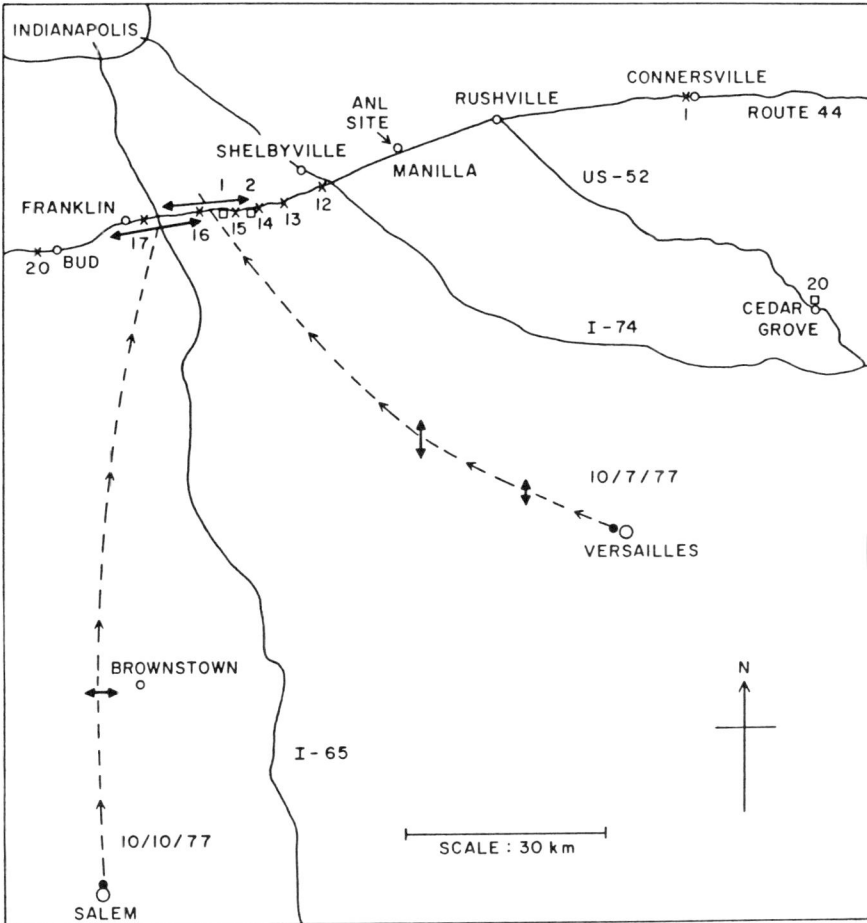

Figure 13.   Location of the SF_6 tracer plume from 2 separate releases. Heavy lines across the plume trajectory represent the aircraft sampling location and the extent of the width of the plume.

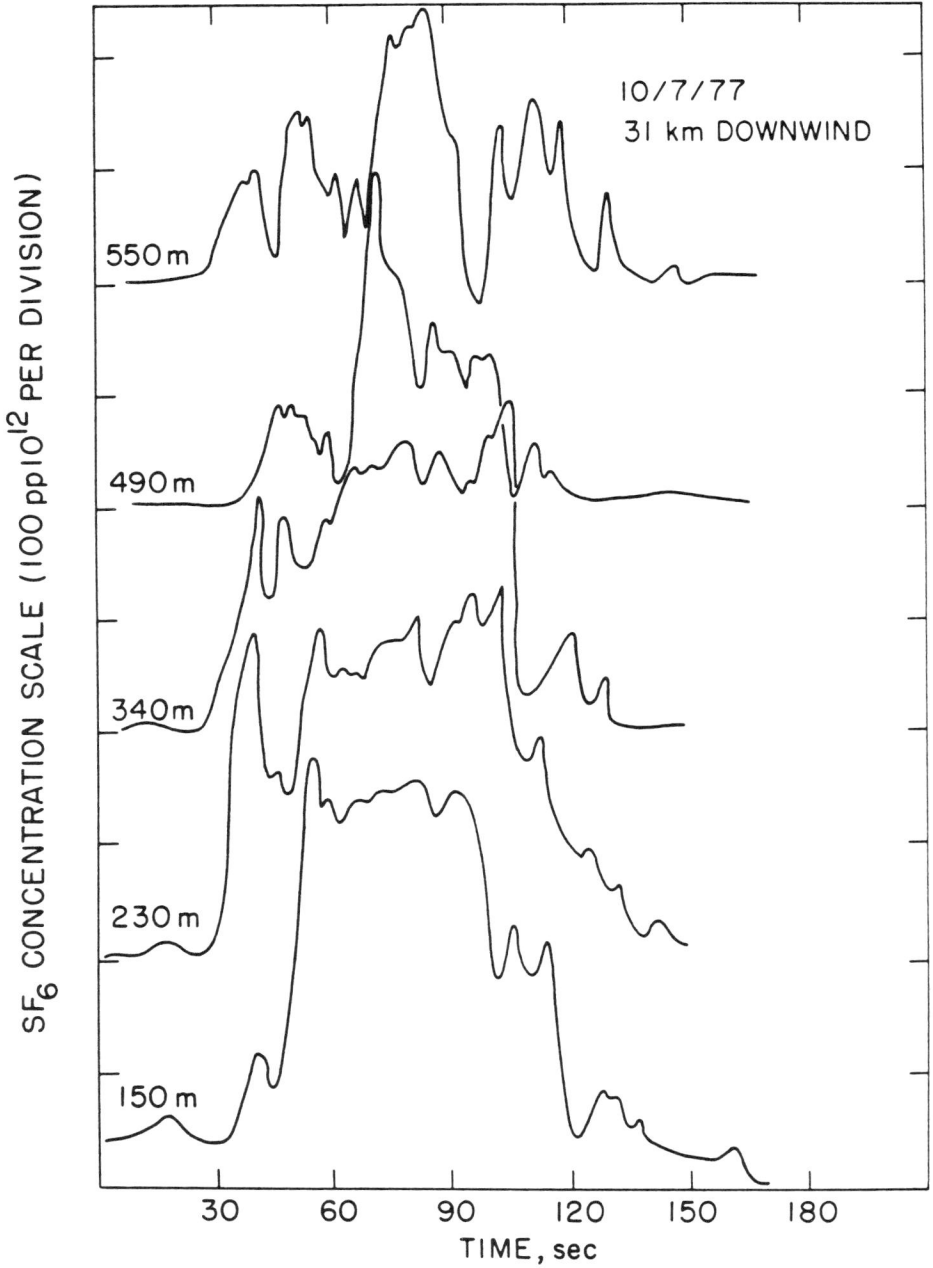

Figure 14.   Staggered plot of SF_6 concentration versus crosswind distance at 31 km downwind at several altitudes.

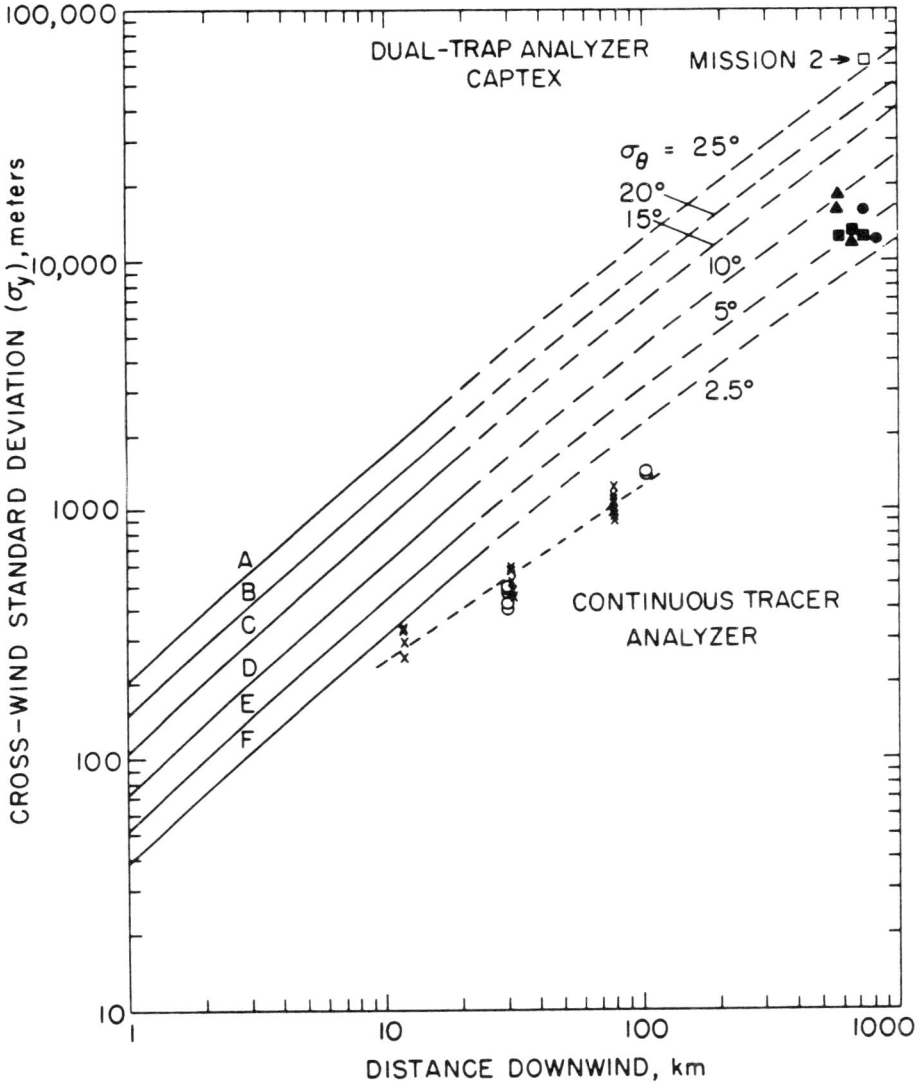

Figure 15.   Results of CTA-measured plume crosswind standard deviations at 3 distances downwind (12 to 78 km) on an extrapolated Pasquill-Guifford plot, also dual-trap analyzer crosswind standard deviations during CAPTEX 1983 at 600 to 825 km downwind.

Figure 16.   Nichrome wire heating element glows red to desorb PFTs from one trap in the dual-trap analyzer while the other is collecting sample.

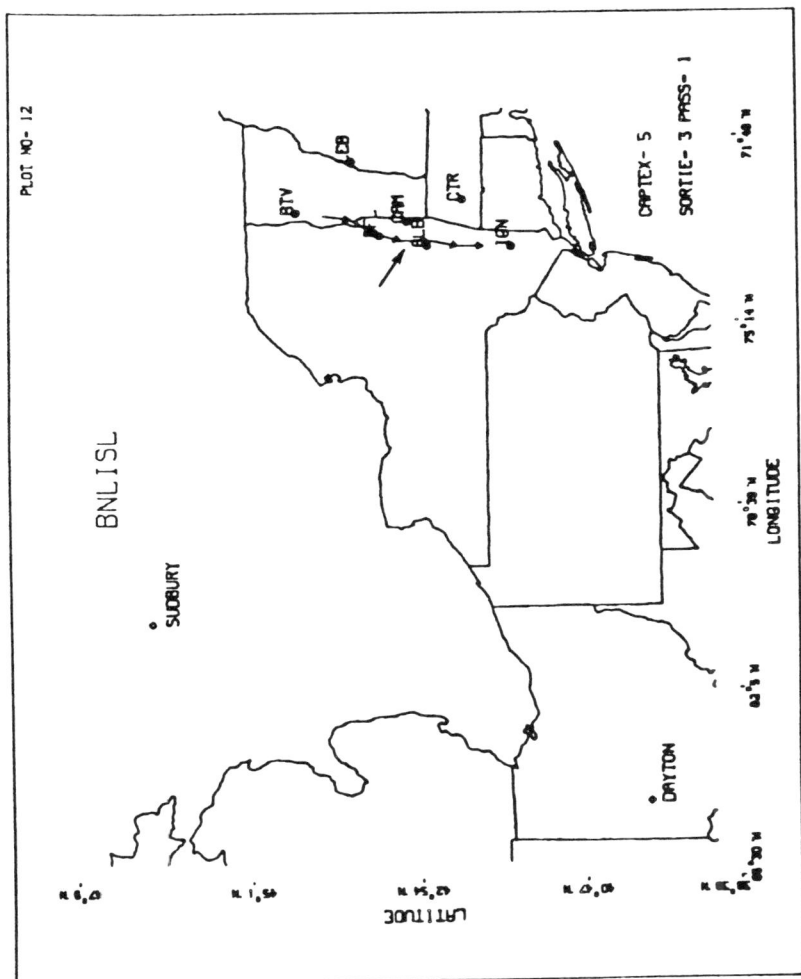

Figure 17.   Map of captex region showing location of dual-trap real-time analyzer on-board aircraft curing traverses at 670 km downwind from the Sudbury release point. Arrowheads on the line represent 10-min sampling intervals for the BATS sampling tubes.

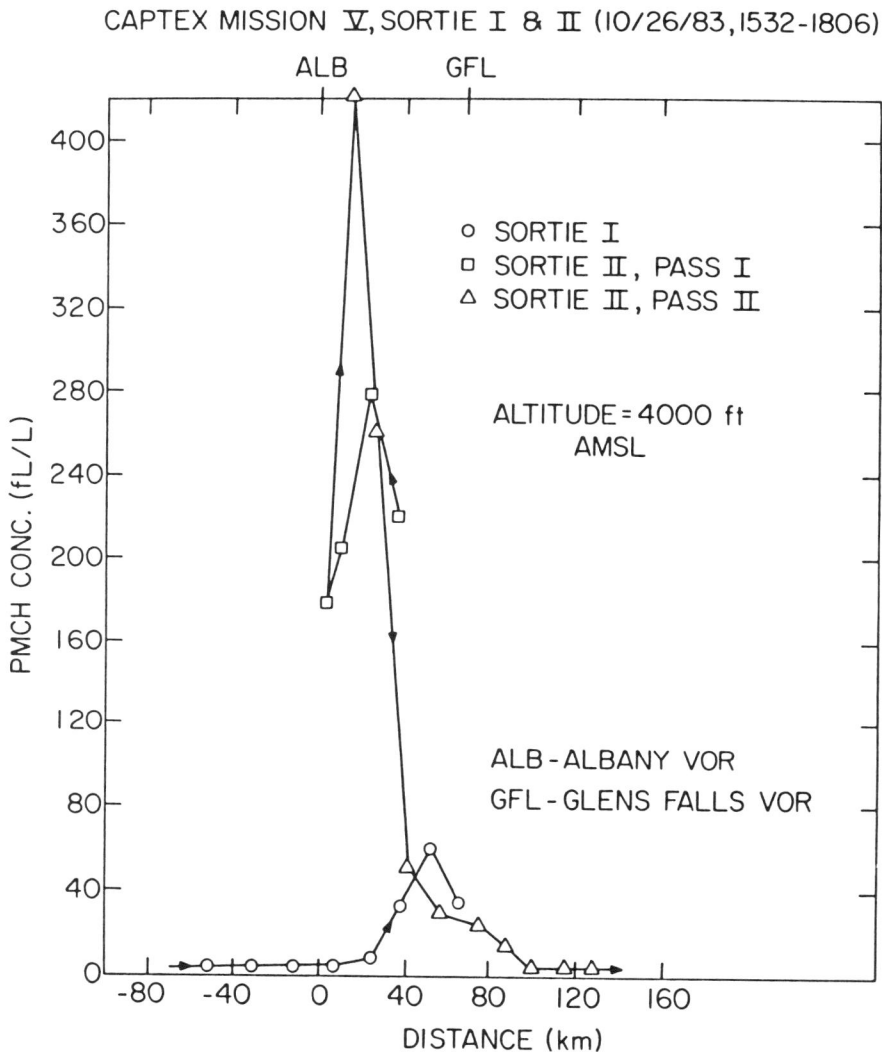

Figure 18.   PMCH concentration from the dual-trap real-time analyzer
during 670 km downwind crosswind traverses showing the arrival
of the plume between Albany and Glens Falls, NY.

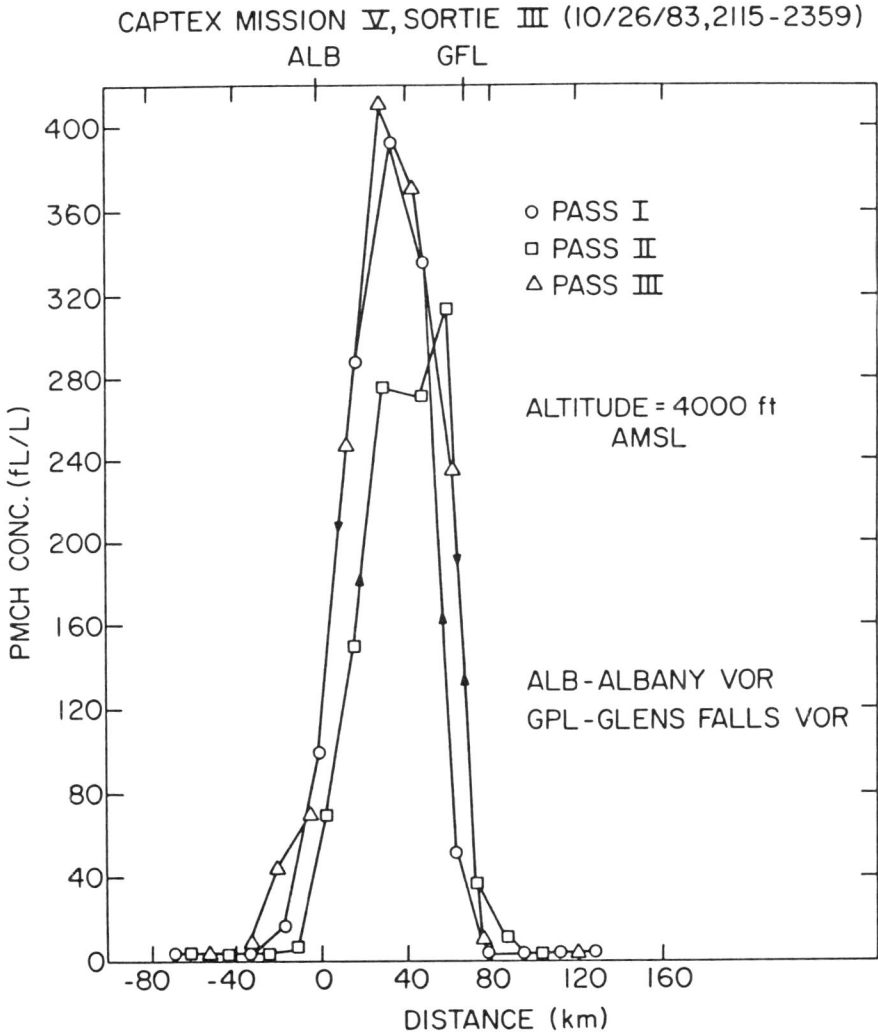

Figure 19.   Subsequent traverses showing the constancy of the plume.

VERTICAL ATMOSPHERIC SAMPLING CABLE
(VASC)

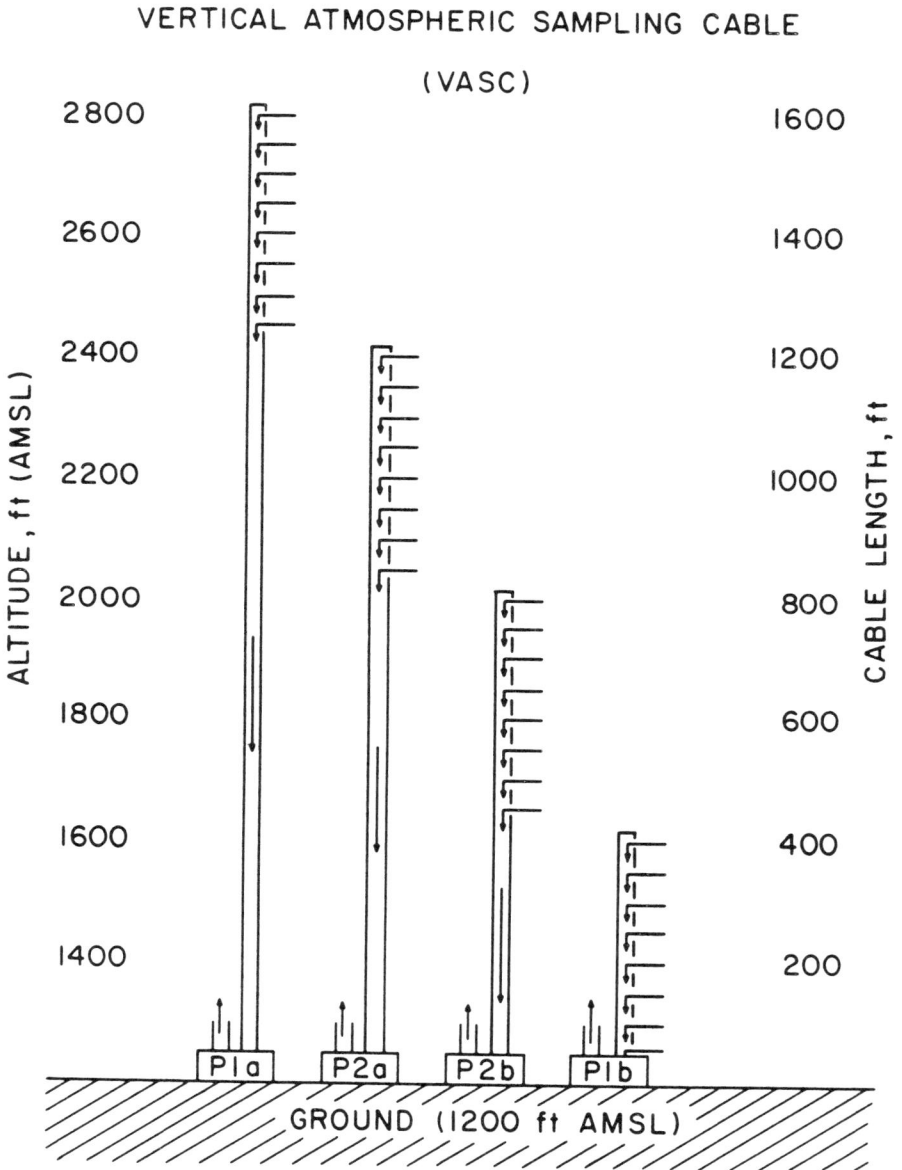

Figure 20.   Schematic configuration of the tethered VASC-I.

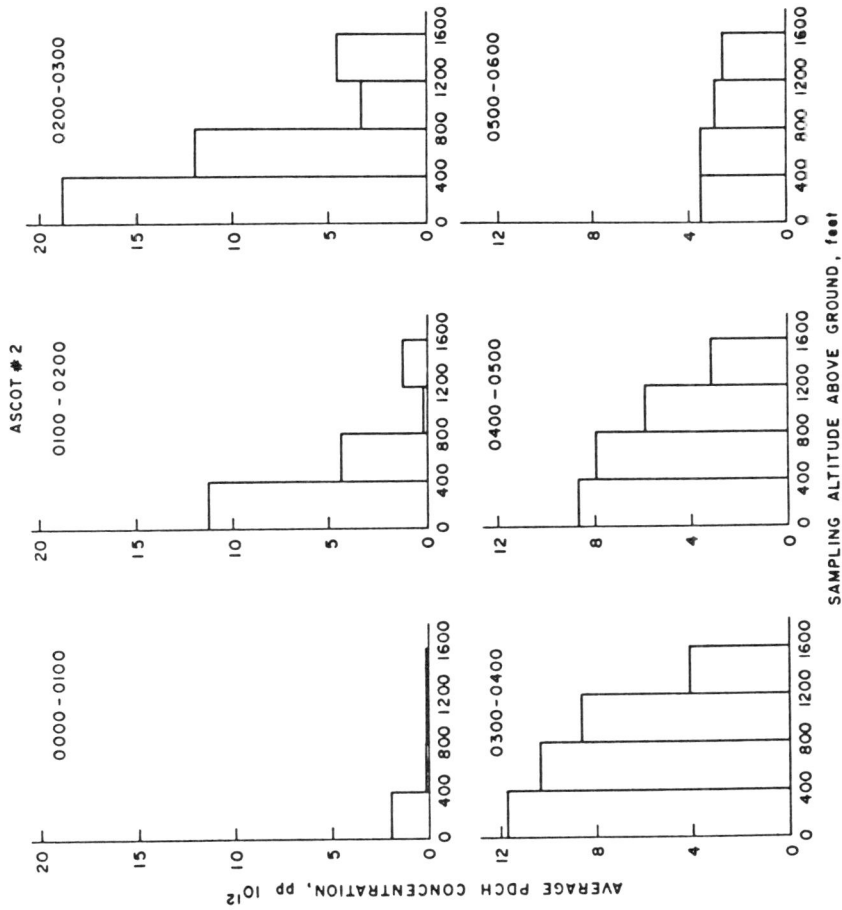

Figure 21.   Vertical distribution of PDCH tracer measured with the dual-trap analyzer connected to the VASC-I.

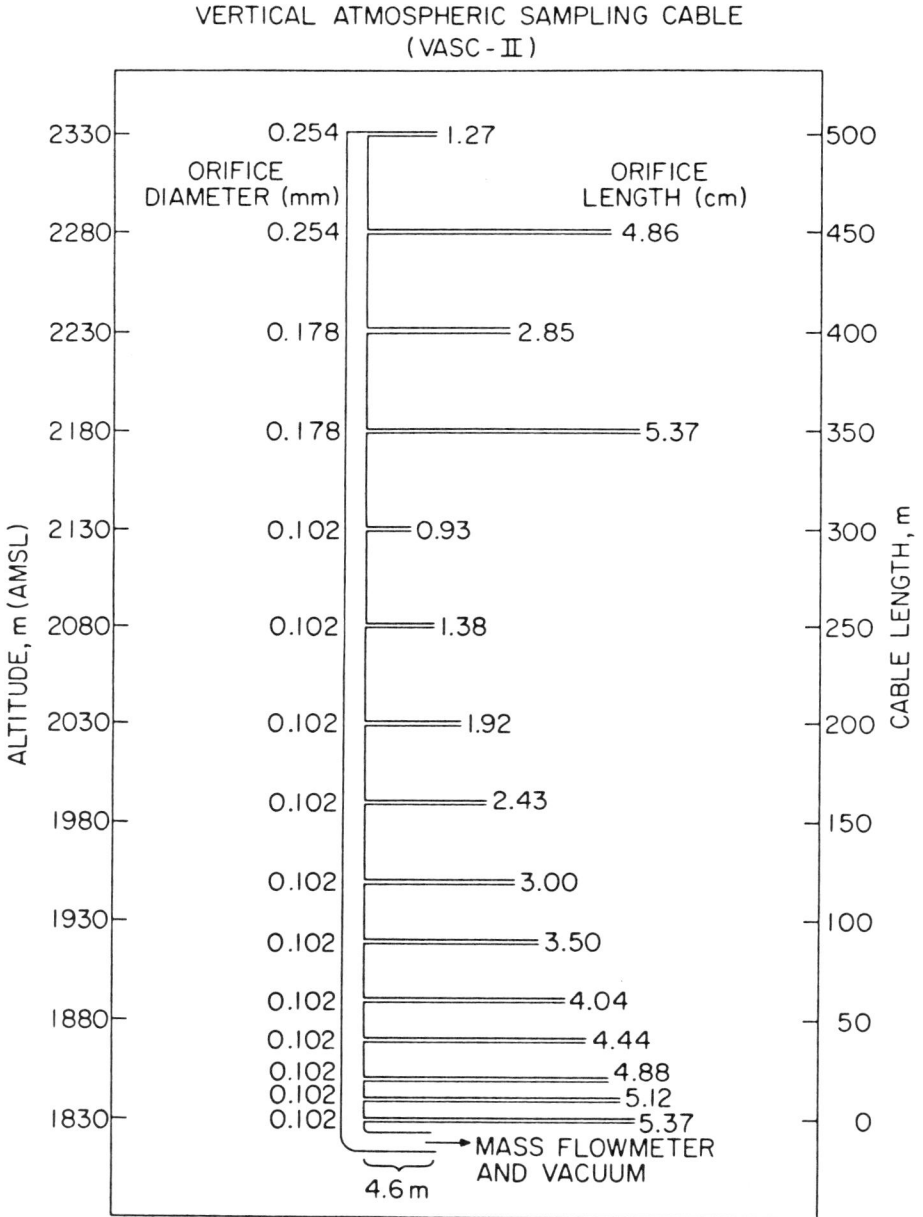

Figure 22.   Schematic representation of the second generation vertical atmospheric sampling cable (VASC-II).

Figure 23.   Programmable PFT sampler (BATS) being analyzed on the automated laboratory electron capture gas chromatograph system.

TABLE 1

PROPERTIES OF GASEOUS CONSERVATIVE TRACERS

Tracer[a]	Symbol	Formula	Molecular Weight	Phase at 20°c	Boiling Point, °c	Supplied Form
Sulfur Hexafluoride	SF_6	SF_6	146	gas	-64	liq. gas
Bromotrifluoromethane	F13B1	$CBrF_3$	149	gas	-58	liq. gas
Perfluorodimethylcyclohexane	PDCH	C_8F_{16}	400	liquid	102	liquid
Dibromodifluoromethane	F12B2	CBr_2F_2	210	liquid	25	liquid
Perfluoromethylcyclohexane	PMCH	C_7F_{14}	350	liquid	76	liquid
Perfluoromethylcyclopentane	PMCP	C_7F_{14}	350	liquid	48	liquid
Perfluorodimethylcyclobutane	PDCB	C_6F_{12}	300	liquid	45	liquid
Deuterated Methane	CD_4	CD_4	20	gas	-160	gas
Deuterated Methane-13	$^{13}CD_4$	$^{13}CD_4$	21	gas	-160	gas

a.   Listed in order of decreasing ambient concentration.

TABLE 2

AMBIENT CONCENTRATIONS AND RELATIVE COSTS[a] OF GASEOUS
CONSERVATIVE TRACERS

Tracer Detection Criteria
Distance: 100 km
Desired concentration: 100 times background at centerline
Release Time: 3 hours

Tracer Symbol	Ambient Conc., fL/L[b]	Cost $/kg	Released[f] Qty., kg.	Relative Tracer Cost 1,000$
SF_6	2000[c]	10	2320	23.2
F13B1	750	15	887	13.3
PDCH	26	120	82	9.8
F12B2	<20	30	<33	<1.0
PMCH	3.6	100	10	1.0
PMCP	2.7	100	9[e]	0.9
PDCB	0.35	500	0.83	0.42
CD_4	0.60[d]	3,000	0.095	0.29
$^{13}CD_4$	0.02[d]	50,000	0.0033	0.17

a.   Does not include cost of sampling and analysis.

b.   1,000 fL/L equals 1 pL/L or 1 part-per-trillion.

c.   Near-urban SF_6 is 2,000 fL/L or more in many locations because of significant use;
     tropospheric background is 850 fL/L.

d.   Values for deuterated methane represent current limits of detection for a 1 m^3 air
     sample; actual backgrounds are about 0.0005 fL/L.

e.   Based on ~85% purity which is readily available.

f.   Released quantity necessary to meet the tracer detection criteria.

TABLE 3

PFT SAMPLER SPECIFICATIONS

Programmable Sampler (BATS)

- 23 sampling tubes in removable lid assembly
- Portable - 1 month internal battery and pump
- 7-day calendar clock
- Programmable start, no. of samples, and sample duration
- Selectable flow rates (0.5 to 40 mL/min)
- Sample recovery by internal thermal desorption
- Size
  - 14 x 10 x 8 in.
  - 7 kg
- Commercially available
  - Complete unit ($5,000)
  - Extra lids ($2,000)

Passive Sampler (CATS)

- Completely passive; no power requirements
- Effective pumping rate (0.14 mL/min)
- Single sample
- Detection limit (0.1 ppt-h)
- 1-month sample
  - Effective sampled volume (6 L of air)
  - Equivalent to 22 fL of background PMCH
- Size (2.5 in. long by 0.25 in. OD; ~5 g)
- Cost ~$3

TABLE 4

AMBIENT PFT DETERMINATIONS WITH PASSIVE SAMPLERS (CATS)[a]
Washington, D.C. area, 30-day period, June 1983

No.	CATS No.	PFT Concentration, fL/L	
		PMCH	PDCH
1	925	3.52	26.3
2	936[b]	22.33	40.1
3	953	3.85	26.9
4	913	4.00	28.5
5	969	3.91	27.9
6	932	3.39	26.3
7	926	3.07	24.3
8	300[b]	23.40	45.8
9	954	3.14	27.3
10	959	2.90	20.5
	Average	3.5+0.4	26.0+2.6

a. Equivalent sampled air volumes were 7.0 and 6.5 L, respectively, for PMCH and PDCH.

b. Excluded from averages; CATS No. 936 had a cracked cap.

TABLE 5

PFT BACKGROUND AIR SAMPLES, YAPHANK, NY
(November 4, 1982)
Ambient Concentration, fL/L

Sample Vol., L	PCDB	PMCP	PMCH	PDCH			
				meta	para	ortho	Total
24.94	0.373	2.76	3.72	12.40	10.30	4.09	26.79
24.36	0.340	2.76	3.57	12.57	10.43	4.41	27.41
24.88	0.359	2.68	3.53	11.44	9.73	3.74	24.91
24.11	0.329	2.69	3.47	10.86	9.34	3.39	23.59
23.37	0.290	2.59	3.50	11.68	9.67	4.13	25.47
22.11	0.383	2.79	3.70	11.61	9.89	4.00	25.50
Avg.:	0.346	2.71	3.58	11.76	9.89	3.96	25.61
	+0.034	+0.07	+0.10	+0.63	+0.41	+0.35	+1.36
Retention: Time[a], min	1.50	1.58	2.71	5.70	6.15	6.75	

a. Retention time is the time from the opening of the Porapak QS trap to the opening of the peak.

TABLE 6

AVAILABLE LONG-RANGE PFTs NAD NEEDED PFT RESEARCH

Available PFTs

Name	Background Conc., fL/L	Cost
1. Perfluoromethylcyclohexane (PMCH)	3.6	$100/kg
2. Perfluoromethylcyclopentane (PMCP)	2.7	$100/kg

Needed PFT Research

- A literature review
- Discussion with PFC[a] manufacturers
- Procurement of samples of existing PFCs
- Contracted synthesis of new or previously available PFCs
- Fabrication and calibration of new PFC standards
- Compatibility with release equipment, samplers, and analyzers
- Field tests
- Cost and purity of PFCs
- Estimated research time – 2 years

a. PFC represents general perfluorocarbon compounds.

TABLE 7

ESTIMATED MATEX YEAR-LONG TRACER NEEDS
(Scale of 1,000 to 1,400 km)

No. of PFT types (1 type per source region)	5
Release duration	4 hours
Release frequency	Every 2-1/2 days
Assumed PFT ambient background	3 fL/L (vol./vol.)
Release rate per source region (for peak concentration 100x background)	50 kg/h
Term of experiment	1 year
Quantity of each PFT	29 metric tons/year
Current global atmos. PMCH	175 metric tons
Estimated annual production	<5 metric tons
Annual PFT Cost	$15 million

TABLE 8

PROGRAMMABLE SAMPLER PROBLEMS AND RESEARCH NEEDS

Problems

- Pump
  - Too many failures (10 to 20%)
  - Changing characteristics
  - Pump rate is manually selected and constant for each tube

- Programming
  - Manually selected duration and quantity
  - Duration identical for each tube

- Date recording
  - Printer tape unit fails
  - Data transfer is manual

- Sampling tube adsorbent
  - No longer commercially avialable

Research Needs

- Improved or new design pump
  - Electronically selectable pumping rates
  - Increased capacity to ~300 mL/min
  - Constant cyclical rate

- Microcomputer controlled operation
  - Individual tube pumping rates
  - Individual start and stop times

- Remote programming and performance evaluation
  - Hand-held plug-in programming unit
  - Capable of telephone jack remote programming and performance evaluation

- Data handling
  - Solid state memory (removable with lid)
  - LCD display of data
  - Capable of remote telephone jack transmission

- Sampler adsorbent studies
  - Evaluate new adsorbents
  - Develop new source of Ambersorb

- Evaluate new prototype and production line units

- Estimated research time - 2.5 years

TABLE 9

LABORATORY GAS CHROMATOGRAPH ANALYSIS OF PFTs

Present Capabilities

- Isothermal column temperature
  - Detection limit is 0.5 to 5 fL (gaseous)
  - Analysis time is 10 min per sample

- Moderate resolution column
  - 3 to 4 PFT resolution

- Automatic analysis for an entire lid assembly
  - 23 tubes in 4 h
  - Peak area data storage via several options

- Sample capacity per week
  - 184 samples per week for an 8-h day

Needed Research

- Improved resolution packed columns
  - Increase analysis speed and no. of PFTs

- Temperature programming of column
  - Increase analysis speed
  - Improve resolution of all PFTs

- Improved catalytic sample processing
  - Reduce sample loss
  - Longer catalyst lifetime

- Performance of new PFTs
  - Column performance and stability in catalytic processor
  - ECD response

- Round-the-clock analyses

- High resolution GC columns
  - Capillary columns
  - Microparticulate packed column

- High resolution GC techniques
  - Pressure balance switching to remove valves
  - Automated "heart cutting"
  - Automated on-column cold trapping

TABLE 10

REAL-TIME DUAL-TRAP ANALYZER

Present Performance

- Maximum sampling rate of 1 L/min

- Dual traps

    - One trap samples air for up to 4 min
    - The other trap is analyzed in 4 min

- Resolves 2 or 3 PFTs in 4 min

- Detection limit of about 1 fL/L

- Internal microprocessor for reporting concentration by the reference tracer method

Additional Studies

- More rapid trap thermal desorption

- Higher resolution columns and techniques

- Commercial supplier is needed

MEASUREMENT OF DRY DEPOSITION AND RESUSPENSION USING TRACERS

J.A. Garland

Environmental & Medical Sciences Division
AERE, Harwell, Oxfordshire OX11 ORA, UK

ABSTRACT

Tracers have made an essential contribution to the development of the understanding of dry deposition and resuspension. Many stages in the development of both topics have depended on the use of tracers selected for particular properties, and numerous studies in the laboratory, in wind tunnels, and in the field have depended on tracers introduced under the control of the experimenter. This is true of work on the dry deposition of gases where studies of selected tracers have allowed aerodynamic and surface effects to be distinguished. The use of particulate tracers of closely controlled size has also allowed the mechanisms for dry deposition of particles to be evaluated. However, some questions relating to resuspension require tracers distributed over such spatial or time scales that the only feasible means of making progress requires that use be made of adventitious tracers, natural, or manmade.

The paper attempts to indicate the special contribution made by both artificial and adventitious tracers.

I. INTRODUCTION

Tracers of various sorts have been used in numerous investigations of the processes that exchange substances between the atmosphere and the surface. The use of tracers introduced in a controlled manner at a selected point in the atmospheresurface system has enabled individual processes to be studied unambiguously, without interference from other processes, and such experiments have played an important part in the development of understanding deposition and resuspension.

Early investigations of deposition processes were prompted by the discovery of radioiodine in crops near the Hanford nuclear site,[1] and it was natural to use ^{131}I to investigate the rate of deposition. (e.g., see Ref. 2). Similarly, when interest in the deposition of sulphur dioxide developed, radioactive and stable isotopes, ^{35}S and ^{34}S, were used as tracers.[3,4,5] Several experiments have employed tracers, radioactive and otherwise, to investigate the deposition and also the resuspension of particulate material. Such tracers have enabled the minute amounts of material transported during a brief experiment to be distinguished from the large mass of material present in the surface in outdoor experiments (see Ref. 6 for a review). Tracers chosen for specific properties have enabled the mechanisms limiting transport to be investigated in laboratory studies, and so have aided the establishment of a framework for the description of surface exchange processes.

Tracers introduced deliberately for specific experiments have been widely used, but some use has also been made of adventitious tracers. The second class of substances often brings the disadvantage that dry deposition may be over-estimated unless wet deposition can be taken into account, or that the resuspended aerosol is mixed with that from other sources. Some reference to adventitious tracers will be made where they have contributed significantly, but generally the accent in succeeding sections will be on tracers applied under the control of the experimenter.

In the succeeding sections, dry deposition and resuspension will be introduced in turn with a brief statement of the conceptual framework used in their description. Subsections will then follow, presenting a description of the contribution of tracers to these fields and an understanding of the present state of knowledge of these topics.

II. DRY DEPOSITION

A. Introduction

Dry deposition comprises several mechanisms that result in the sorption and retention of particulate or gaseous material from the air by the surface, be it dry or wet. Many substances are present in the atmosphere at such low concentrations that the mechanisms that transport them to the surface are not influenced by the quantities of the tracer present. Then the deposition flux is proportional to the concentration and it is advantageous to express results in terms of the deposition velocity

$$v_g = \frac{\text{flux to the surface, } F}{\text{Concentration } X} \,.$$ (1)

The concentration usually varies slowly with height above the surface. For precision, the reference height z_r at which X is measured should be specified; in practice the variation of X, and therefore v_g, with z_r is usually small compared with the variability of v_g due to other causes. In this paper, F is usually taken to be the flux to the entire plant canopy, soil, or other kind of surface per unit ground area. Occasionally elsewhere, F is taken to be the flux per unit leaf area.

Dry deposition is taken to describe the transport of material from a height z_r within the turbulent boundary layer to the surface. It can occur only if there is both a mechanism for capture at the surface and for transport through the air to the surface. In the boundary layer, molecular processes are normally dominated by turbulence. Gases and small particles presumably follow turbulent eddies in the same way as heat and water vapor. Momentum is also conveyed by eddies, with some moderate differences in rate, depending on atmospheric stability. Transport of heat, water vapor, and momentum have been the subject of careful investigations by micrometeorologists and may be treated as tracers for gas and particle transport, facilitating estimates of the rate of transport to the surface in somewhat ideal circumstances.

The analogy between gas (or particle) transport and that for heat or water vapor fails in the close vicinity of the leaves, stems, and other roughness elements clothing the surface. These surfaces are surrounded by slowly moving air, where molecular processes (or mechanisms for conveying particles through stagnant air) become important.

Finally, at the surface, the rate of chemical reaction or adhesion influences the deposition rate.

The influence of these three processes, acting in series, on the deposition velocity v_g is often expressed in terms of an electrical analogy:

$$v_g^{-1} = r = ra + {}^r b = rs, \qquad (2)$$

where r is called the total resistance, and r_a, r_b, and r_s are partial resistances representing, respectively, transport through the turbulent boundary layer, the viscous surface layer, and the surface sorption.

The resistance for the turbulent boundary layer, often called the aerodynamic resistance, can be predicted from micrometeorological studies. Experimental studies of the deposition of tracers of known surface properties have allowed estimates of r_b to be made. For gases, the behaviour of r_s can sometimes be deduced from experiment, so that all three resistance components, and hence v_g, can be predicted. An immediate consequence of Eq. (2) is that the deposition velocity cannot exceed $(r_a + r_b)^{-1}$, even for the most reactive gases. Equation (2) is useful for the analysis of the deposition of gases and of particles of diameter up to about 0.1 μm. Brownian diffusion transports such particles through the laminar boundary layer, as molecular diffusion does gases.

Much larger particles deposit chiefly by interception, impaction, and sedimentation. The inertial trajectory of these particles may carry them through the boundary layers, bypassing r_b and, partly, r_a. A modified form of Eq. (2) may be appropriate.

B. Use of Tracers for the Dry Deposition of Gases

The formulation of Eq. (2) encourages separate investigation of the partial resistances.

Exchanges of heat, water vapor, CO_2, and momentum have been studied extensively by micrometeorologists and are the tracers which have contributed to the understanding of r_a (e.g., Refs. 7,8,9). r_a is most conveniently evaluated by measuring velocity profiles and considering momentum exchange. A significant difference between the transfer of momentum and that of heat and water vapor occurs when the surface is warmed by the sun, so that density effects influence turbulent exchange. Evidence is mixed, but it is usual to accept that vapors are exchanged at the same rate as heat. (The difference is rarely as large as other uncertainties in the analysis of deposition.) The most convenient way to evaluate r_a is often to measure velocity profiles and correct the resistance for momentum exchange for the difference between momentum and heat. Formulations for this procedure were published by Garland and Wesely (Refs. 10,11), among others.

A further difference between the exchange of momentum and that of heat and vapor occurs in the immediate vicinity of the surface, where molecular diffusion D and viscosity ? become important, and where pressure forces acting on roughness elements take part in the transport of momentum. The partial resistance r_b is introduced to accommodate this difference. r_b is expected to depend on the Schmidt number (?/D) and is greater for gases of high molecular weight. Early measurements were performed in a wind tunnel using the evaporation rate of camphor to give a measure of r_b (Ref. 12). Studies by Chamberlain[13,14] involved measurement of evaporation of water from fully wetted surfaces and also the deposition of ^{212}Pb vapor, formed when ^{220}Rn (thorium emanation or thoron) undergoes

radioactive decay in nucleus-free air. Uptake of radioactive iodine tracers at
very reactive surfaces has also been studied by Chamberlain et al.,[15] and
field data for heat transfer from vegetation has been interpreted to give
estimates of r_b (Ref. 16). The key feature of all these measurements is a
knowledge of the concentration X_o at the surface. For camphor and water,
well-coated surfaces exert the saturation vapor pressure, whereas for I_2 and
^{212}Pb, in the conditions of the measurements, $X_o = 0$.

Such correlations allow adequate estimates of r_b for interpretation of
other measurements. Deposition of SO_2, O_3, PAN, HNO_3, HF, NO_2, and
other reactive gases has been measured in numerous field and laboratory
studies. Many of these measurements were reviewed recently.[17] With
sufficient micrometeorological measurements, r_s can be reduced, and an
understanding of the variation of r_s for each gas-surface pair can be built up.
In principle, this allows deposition velocities to be predicted for a wide range
of conditions; in practice, sufficient information is available for SO_2, O_3 to
provide only a general indication, and uncertainties up to a factor of 2 must
be admitted. For other gases the situation is even less satisfactory.

Special mention must be made of isotopic tracers used in such
investigations. ^{35}S and ^{34}S have been used[3,4,5] to investigate particular
features of SO_2 uptake by foliage (e.g., see Fig. 1). These enable S collected
during a period of minutes to be distinguished from that accumulated during
the entire growth of the plant. They also permit investigation of the site of
uptake and use of the deposited S by the plant, (e.g., by autoradiography[18])
but little use has been made of this facility. Few such convenient isotopes
are available for other gases. ^{15}N has been used in investigation of NH_3
uptake[19] but has not, to the author's knowledge, been applied for studies of
nitrogen oxide exchange.

For some gases, uptake may be reversible or partly reversible. In such
a case the use of an isotopic tracer in a brief exposure

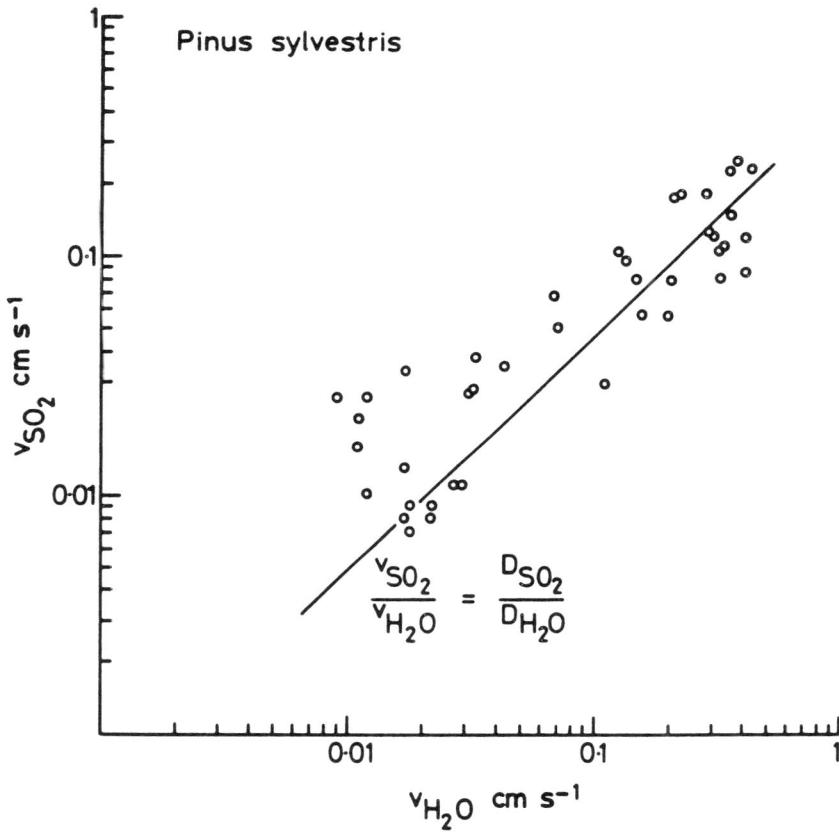

Fig 1.   Uptake of SO_2 by pine needles, measured using [35]S (Ref. 5). The relationship between the deposition velocity V_{SO_2} (based on projected area of needles) and the analogous velocity for transpiration of water vapor V_{H_2O} shows that the same limiting mechanism (diffusion via stomata) controls both.

may give the total flux rather than the net flux. An example is the use of tritiated water vapor (HTO) to trace water vapor exchange; HTO uptake from air to plant may be recorded even when vegetation is transpiring.[20] Similarly, $^{14}CO_2$ exposure indicates total photosynthesis because, in a brief exposure, the ^{14}C does not become available for respiration. ^{15}N, used to label NO_x or NH_3, might overestimate the net flux if there are significant return fluxes of NO and NH_3 from soil or vegetation. Although some emission of reduced S may occur, it is unlikely that this is significant in comparison with SO_2 deposition in industrial regions, and S isotope methods probably do not give significant errors of this kind.

C. Tracers Used in Particulate Dry Deposition Studies

Measurements of the dry deposition velocity of particles can be categorised as follows:

(1)   wind tunnel measurements using artificial tracers

(2)   field measurements using artificial tracers

(3)   field measurements using adventitious tracers

Wind tunnel experiments and field measurements using artificial tracers have enabled effects of particle size, wind speed, and surface texture to be investigated.

Tracers used have included Lycopodium and other spores and pollens (often made radioactive[21,22] or dyed[23] to facilitate detection) and a variety of synthetic monodisperse particles (e.g., see Refs. 21 and 24-28). Radioactive tracers have particular advantages because they can often be measured without separation from the collecting surface, so that problems associated with ensuring quantitative separation do not arise.

Data obtained in wind tunnel studies are represented by symbols and the solid curve shown in Fig. 2. The variation of deposition velocity with particle size can be explained

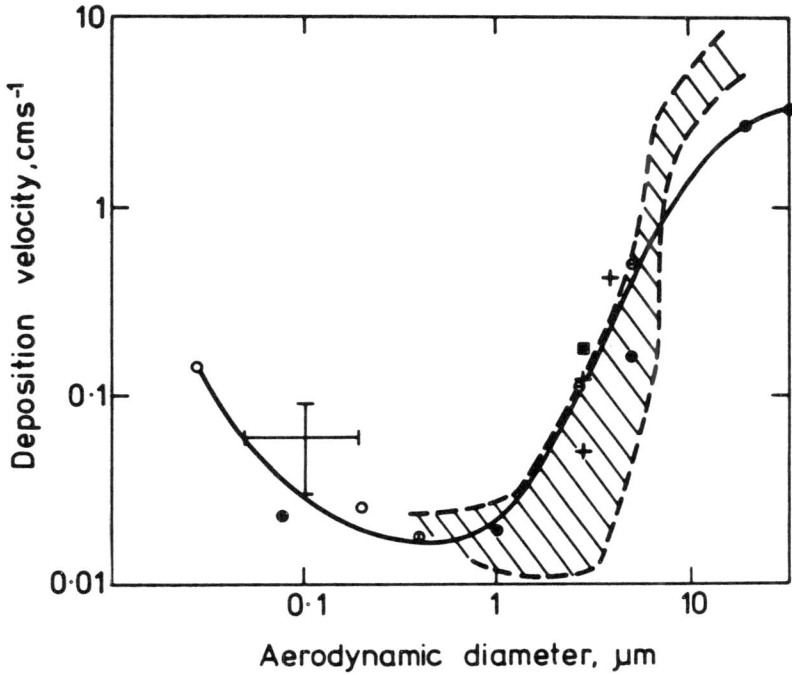

Fig 2.   A comparison of some measurements of deposition velocity to
            grass, measured using artificial tracers of controlled size
            distribution in the wind tunnel and in field experiments.

Wind tunnel data
·   Chamberlain[21]: Spores, pollen, Aitken nuclei, etc. [212]Pb label.
·   Little and Wiffen[58]: motor exhaust lead.
·   Clough[57]: spores, Aitken nuclei, ZnS, and oleic acid.
·   Little[26]: Polystyrene spheres labeled with [99m]Tc.
·   Garland[51]: iron oxide particles labeled with [59]Fe.

Field data
Crosses: Garland[51], using iron oxide spheres labeled with [59]Fe.
Hatched area: Jonas and Vogt[24] using copper sulphate labeled with [64]Cu.

theoretically. Results obtained by other workers[25,27,28] yield curves of
similar shape and magnitude for surfaces ranging from water to forest.
Figure 2 indicates that field studies using artificial tracers of controlled size
are consistent. The field study results include data from 52 experiments[24] in
which radioactive copper sulphate particles were used. Slinn et al.[29] used
submicron particles of indium, dysprosium, and europium as tracers. These
elements can be estimated with high sensitivity by neutron activation
analysis. Despite some unexplained variability, the results were mostly
broadly consistent with Fig. 2.

In all the experiments introduced in the previous paragraph, the
deposited material was measured directly in samples of the receiving surface.
A further class of experiments depends on measuring the depletion of a plume
of tracer particles as it is carried downwind. The difficulties of allowing for
the fall in concentration due to dilution are substantial. Attempts to use the
sea-salt aerosol travelling across North America illustrate this problem.[29] In
an attempt to avoid the dilution problem, two tracers are released
simultaneously. One tracer is a stable gas, which will not undergo deposition,
whereas the second is the particulate material under test. Changes in the
ratio of concentration of the particulate and gaseous tracers enable depletion
of the plume by deposition to be estimated. In an early experiment of this
kind,[30] ^{133}Xe and 1 to 5 µm zinc-cadmium sulphide were used. The loss of
particles between ranges of approximately 16 and 60 km would have
corresponded to a deposition velocity of 15 cm s^{-1}, but the authors noted a
decrease in the fluorescent brightness of individual particles with travel time
and suggest that some process other than deposition was responsible.
Sehmel[31] used SF_6 and a lithium carbonate-aerosol to determine particle
depletion from nocturnal density-driven air flows in canyons. The lithium-
containing particles were counted individually by observing light emitted
from a hydrogen flame. The plume depletion at 1.5 m above the surface,
between sampling points at ~2.5 km downwind of the source, ranged from 42
to 74%. In another series of experiments,

the average depletion below 15 m varied from 1 to 11%. Wind speeds in these drainage flows were light, about 0.5 to 3 m s^{-1}. The author declined to deduce a deposition velocity, but it seems that values of a few millimetres per second would explain the observed depletion. Such values are an order of magnitude larger than the expected value for particles of the size range (0.3 to 1.5 μm) studied in the experiment.

A number of methods have been applied to estimating the dry deposition rate of tracers of opportunity. Deposition in rain often dominates this process, but deposition to surfaces mounted below rain shelters can readily be measured. The deposit on a suitably mounted collector may be compared with the concentration in air to deduce the deposition velocity. Particulate tracers generally have wide size distributions. Cawse[32] gives deposition measurements to a filter paper surface in the field and measurements of size distribution determined with an Andersen impactor, and Clough[33] studied the performance of the same collector in a wind tunnel. Thus, a comparison of observed deposition velocity and that expected from the wind tunnel study is possible. The results (Fig. 3) indicate that deposition in the field is several times higher than expected. A number of other studies point to the same conclusion. Several factors may contribute to the discrepancy. Some of the tracers have volatile forms in the atmosphere. The size measurements do not give complete information for particles above about 10 μm in diameter, and particles larger than a few microns in diameter may not be sampled quantitatively in the concentration samplers. In addition, the exposure of the filter paper pad in the field, at 1.5 m under a rain shield, differences in turbulence between the field and wind tunnel, and the wide range of wind speeds in the field complicate the comparison.

Other methods include the measurement of concentration gradient with height and eddy correlation. For the former, rather small concentration differences must be measured over a range of

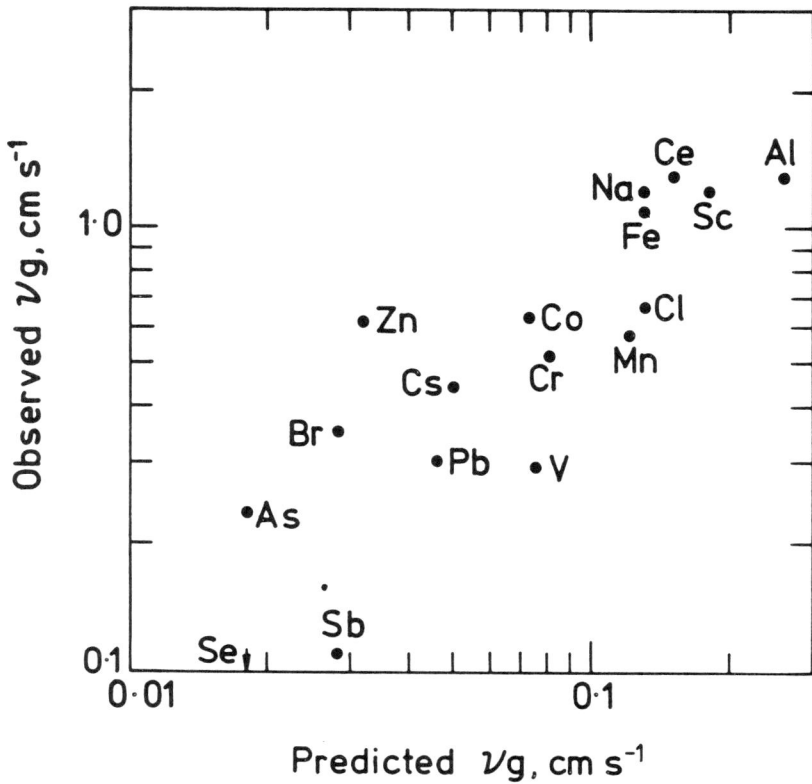

Fig 3.   Deposition velocities for adventitious trace elements to filter paper
exposed under a rain shelter. The observed deposition velocity is
consistently an order of magnitude larger than the value predicted
from measured size distributions and wind tunnel measurements of
deposition to the same surface.

heights within the lowest few meters of the atmosphere. Doran and Droppo[34] illustrate the difficulties that arise in practice and suggest that sometimes the deposition velocity of sulphate reaches 0.3 or 0.4 cm s^{-1}. Davies and Nicholson[35] registered over a hundred measurements of sulphate deposition to grass. The spread of values (from -0.8 to + 1.2 cm s^{-1}) reflects random errors of measurement as well as variability of v_g. Median values of order 0.1 cm s^{-1} were obtained, which is roughly consistent with wind tunnel results.

Eddy correlation requires the measurement of concentration (or some linearly related property) with a time resolution of order 0.1 s. For such purposes, the "tracer" may be a chemical component of the aerosol (e.g., sulphur, Ref. 36) or an electrical or optical property.[37,38,39] Such methods have resulted in a range of results, including some periods of upward fluxes and sometimes large downward fluxes implying deposition velocities of around 1 cm s^{-1}. It is difficult to reconcile these results with those in Fig. 2, and there would be some benefit in a careful review of the data and the precision and sensitivity of the various methods to a range of environmental parameters.

The fate of deposited material is often important. In environmental considerations for the nuclear industry, the retention of deposited activity on crops and transport through food chains is of importance in determining the resulting ingestion dose. Such considerations are also important for chemical toxins such as lead. Investigations have used radioactive tracers to determine the retention time.[40,41] In a large-scale test in Italy,[42] the lead tetraethyl added to petrol over a period of about 5 years was made from ore with an unusual isotope ratio. The change in isotope composition in various materials allowed the contribution of petrol-lead to blood-lead to be observed and furnished information on the mechanisms of transfer. Examples of the results are shown in Table I.

III. TRACERS IN RESUSPENSION STUDIES

A. Definitions

Processes that lift material from the surface so that it becomes suspended in the atmosphere can be described properly as suspension. The suspension of material previously deposited from the atmosphere is, logically, resuspension. However, the distinction is rarely made, and here we shall treat the terms as synonymous.

Processes operating in the outdoor environment are considered here. Raising of particles by the wind has received most attention, although several measurements of the effect of disturbance by walking, clearing debris, agricultural operations, and traffic are also reported in the literature.

A number of terms have been used to describe observations of resuspension. The resuspension factor

$$K = \frac{\text{concentration in the air } (\mu g \; m\text{-}3)}{\text{surface deposit } (\mu m \; m\text{-}2)}$$

has been much used, but where the surface deposit is nonuniform the value of K becomes dependant on the location chosen for sampling. Thus, the resuspension rate

= the fraction of the surface deposit resuspended per second

has sometimes been preferred, although its measurement in the field can be very difficult.

B. Resuspension Measurements Using Tracers

A number of categories of tracers feature in the resuspension literature:

(1)   adventitious tracers of limited distribution,

(2)   tracers applied in a controlled manner for specific experiments, and

(3)   widespread tracers generally present in soil or sea.

Many early measurements, summarized by Stewart,[43] made use of fall-out deposited close to the sites of nuclear weapons trials and belong to the first category. Values of K span 4 orders of magnitude. In quiescent conditions, a value of 10^{-6} m^{-1} was suggested as representative of the results, with an increase of a factor of 10 when the surface was subject to moderate disturbance. More recent surveys of the resuspension of plutonium at Savannah River, Rocky Flats, and Hanford have been stimulated by the high sensitivity of detection methods and the radiotoxicity of plutonium. Sehmel[44] reviewed this data and noted that although effects of wind speed, sampling height, and particle size could be distinguished, the data were "too complex to develop models for resuspension."

A similar situation can be achieved by using deliberately dispersed tracers (category 2). Stewart[43] quotes results obtained with yttrium chloride (labeled with ^{91}Y), polonium, and uranium oxide. The results of some experiments showed a steady decline of K over a few days. Sehmel[45] used calcium molybdate in a 5-year resuspension experiment at Hanford. No systematic variation with weathering time was observed. Two reports[46,47] detail measurements of resuspension of radioactive tracers ($^{185}WO_3$, $^{59}FeCl_3$ in solution, and silt and spherical iron oxide particles labeled with ^{59}Fe) from grass and soil surfaces. A wind tunnel was used to provide a degree of environmental control. This device was constructed on a field so that the surface of the ground formed the lower boundary of the tunnel. Results indicated consistent variations of resuspension factor with wind speed, time of exposure to wind, and some indication of variation with particle size.

A limited number of publications discuss the effects of mechanical disturbance on outdoor, resuspended concentrations. Sehmel[48] used fluorescent ZnS to measure resuspension from a roadway. Hodgin[49] observed resuspension of plutonium from roads at Rocky Flats. Other workers[50] observed an increase in the concentration of plutonium in air during cultivation of contaminated soil at Savannah River Plant. These studies indicate that a large increase in resuspension occurs under the influence of traffic or other mechanical disturbance. A single pass of a vehicle may suspend ~10^{-2} of the surface dust from a traffic lane on a road surface.[47]

Most of the studies discussed above relate to tracers distributed on a scale measured in meters to a few kilometers for periods of days to a few years. Extrapolation to older deposits of greater extent involves a number of questionable assumptions. The environment contains numerous widely distributed tracers of great age, and some attempts have been made to deduce resuspension rates for these tracers. Nuclear weapons testing peaked in 1963, and most of the weapons fallout has been on the ground for about 2 decades. Continued, sporadic testing contributes substantially to current air concentrations, but consideration of the concentration in air and the deposit in soil in 1980 gave an upper limit to the resuspension factor of ~2×10^{-10} m^{-1} for Great Britain.[51] Similarly, the ratios of concentration of several trace elements in air and soil (including Al, Sc, Fe, Mn, Ce, and Th) indicate airborne concentration of about 7 µg m^{-3} of soil. It has been suggested that about 40% of this material is explained by industrial emissions. Then a value of about 4×10^{-11} m^{-1} for K may be appropriate for a tracer with a mean depth (like the ^{137}Cs from fallout) of 5 cm in soil.[51]

Adventitious tracers have also been used to observe suspension of material from the sea surface. Not surprisingly, the major constituents of sea salt are enhanced in rain and deposition samples at coastal and sea-surface sites, in comparison to samples

collected inland. This gives a measure of the amount of sea spray that
contributes to the samples. However, the enhancement for several trace
elements is much greater than expected from the concentration of the
element in sea water and the quantity of sea spray represented by the NaCl
in sample. That is, the concentration factor

$$CF = \frac{R \text{ in sample}}{R \text{ in sea water}},$$

where the element ratio

$$R = \frac{\text{concentration of element X}}{\text{concentration of sodium}}$$

exceeds 1 by several orders of magnitude for some elements (including Al, Sc,
Cr, Fe, Co, Ni, Cu, Zn, As, Pb).[52] Several techniques have demonstrated
that CF 1 for samples of the seasurface microlayer (a layer collected from
the sea surface, the thickness depending on the technique: generally 300 μm),
and it is natural to suppose that this microlayer is the source of spray.
Pattenden et al.[53] and Piotrowicz,[54] among others, have demonstrated that
spray generated by bubbling air through sea water in situ shows large
enrichments.

Where contaminants are discharged to sea, transfer from sea via the
atmosphere to land may result. This process is illustrated by the behaviour of
certain radioisotopes discharged via a pipeline to the Irish Sea from the
nuclear plant at Sellafield, Cumbria, in North West England. Soil surveys and
air sampling along the coastline have demonstrated that some of the
discharged plutonium isotopes and [241]Am is returned to land. A survey in
1977, over 20 years after discharge to sea began, showed that only about 10^{-4}
of the total Pu discharged had returned to land and was distributed along a
40-km stretch of coast.[55] The mean age of the plutonium discharge at that
time was 7 years. Collection of sea spray by

cloth screen collectors[55,56] has demonstrated a large enrichment effect for Pu and Am, both of which attach to silt in the turbid coastal waters. The ^{137}Cs remains in solution, does not show a substantial enrichment, and is not significantly transferred to land.

IV. FINAL REMARKS

The preceeding sections show that measurements using tracers, adventitious and artificial, have contributed an essential part of our understanding of the processes of deposition and resuspension.

A wide range of tracers have been used (Table II). For studying gaseous deposition, it is usually necessary to use a tracer of the required chemical properties, so the choice is limited to isotopes of the component elements of the gas. There may be several isotopes available (e.g., the 12 radioactive isotopes of iodine with half-lives exceeding 1 hour) or very few (for sulphur there are four stable isotopes and only one radioactive isotope of suitable half-life, whereas for nitrogen only the rare, stable ^{15}N is of use).

For particle studies, the choice is wider because the physical characteristics (size, density, shape, and parameters likely to influence adhesion) usually dominate behaviour. Many tracers have been used, and it is worth reviewing the properties of some of the leading types.

Radioactive tracers have a number of advantages. With appropriate choice, the background in the environment may be negligible. Very high sensitivity of detection can be achieved (for gamma emitters, usually ~1 Bq can be measured readily; this may be equivalent to as little as 10^{-17} g of the isotope). Gamma-emitting isotopes can often be measured without separation from the sample matrix, whether it is vegetation, soil, or filter paper. If an isotope of short half-life is chosen, repeated experiments can be performed without contamination from early experiments confusing the results of subsequent ones. Unfortunately, the radiological hazards perceived to be associated with radioactive tracers have caused their application to be restricted, and their use in environmental studies is effectively banned in some countries. Although control is undoubtedly necessary, such an indiscriminate limitation seems inappropriate because careful choice of the isotope can allow experiments involving extremely small radiation doses to operator and public alike.

A range of other methods of analysis offer high sensitivity for suitable tracers. Neutron activation analysis allows measurement of subnanogram quantities of many trace elements, some of which (dysprosium, indium, europium, irridium, lutetium, gold, hafnium ...[29]) are rare elements with low background concentrations in most environmental samples. Many metals can be analysed at levels of a few picograms by various spectroscopy techniques, proton-induced x-ray emission, anodic-stripping voltammetry, etc.

Mass spectrometry offers the possibility of using isotope ratios to distinguish tracer from background. For the lead study mentioned previously, as little as 10^{-8} g of Pb per sample allowed the isotope ratio to be determined to about 1 part in 10^3. Plasma source mass spectrometry allows many trace elements to be measured simultaneously. Reviews in the analytical literature may help choose tracers and measurement methods.

With such a wide range of techniques available, allowing the use of tracers with a wide range of properties, there should be no difficulty identifying potential tracers for future studies.

REFERENCES

1.   Parker H. M. (1956) Proc. Int. Conf. on Peaceful Uses of Atomic Energy 13, 360363.

2.   Chamberlain A.C. and Chadwick R.C. (1963) Nucleonics 11, 22-25.

3.   Garland J. A., Clough W. J. and Fowler D. (1973) Nature 242, 256-257.

4.   Belot Y., Bourreau J.C., Dubois M. L. and Pauly C. S. (1974) FAO/IAEA: Isotope ratios as pollutant source and behaviour indicators. IAEA-SM-191-18.

5.   Garland J. A. and Branson J. R. (1977) Tellus 29, 445-454.

6.   Sehmel G. A. (1980) Atmospheric Environment 14, 983-1011.

7.   Businger J. A., Wyngaard J. C., Izumi Y. and Bradley E. F. (1971) J. Atmospheric Sciences 28, 181-189.

8.   Dryer A. J. and Hicks B. B. (1970) Quart J. R. Met. Soc. 94, 318-332.

9.   Pruitt W. O., Morgan D. L. and Lourence F. J. (1973) Quart J. R. Met. Soc. 99, 370-386.

10.   Garland J. A. (1977) Proc. R. Soc. Lond A 354, 245-268.

11.   Wesely M. L. (1983) Chapter 8 of Trace Atmospheric Constituents. Ed. S. Schwartz. John Wiley and Sons Inc.

12.   Owen R. R. And Thomson W. R. (1963) J. Fluid Mech. 15, 321-324.

13.   Chamberlain A. C. (1966) Proc. R. Soc. London A 290, 236-265.

14.   Chamberlain A. C. (1968) Quart J. R. Met. Soc. 94, 318-332.

15.   Chamberlain A. C., Garland J. A. and Wells A. C. (1984) to be published.

16.   Garratt and Hicks B. B. (1973) Quartz J. R. Met. Soc. 99, 680-687.

17.   Garland J. A. (1983) VDI Bericht 500, Verein Deutscher Ingenieur, Berlin.

18.   Fowler D. (1976) PhD Thesis. University of Nottingham.

19.   Porter L. K., Viets F. G. and Hutchinson G. L. (1972) Science 175, 759-761.

20.   Garland J. A. (1980) Water Air and Soil Pollut. 13, 317-333.

21.   Chamberlain A. C. (1967) Proc. R. Soc. London A 296, 45-70.

22.   Chamberlain A. C. and Chadwick R. C. (1972) Ann. Appl. Biol. 71, 141-158.

23. Aylor D. E. (1975) J. Appl. Meteorol. 14.

24. Jonas R. and Vogt K. J. (1982) Jul-1780. Kernforschungsanlage, Julich.

25. Moller U. and Schumann G. (1970) J Geophys. Res. 75, 3013-3019.

26. Little P. (1977) Environ. Pollut. 12, 293-305.

27. Sehmel G. A. and Hodgson W. J. (1978) PNL-SA-6721, Battelle Pacific Northwest Laboratory, Richland WA.

28. Belot Y. and Gauthier D. (1975) Heat and Mass Transfer in the Bio-sphere. Part I: Transfer Processes in the Plant Environment, Halstead Press, New York. 583-591.

29. Slinn W. G. N., Katen P. C., Wolf M. A., Loveland W. D., Radke L. F., Miller E. L., Ghannam L. J., Reynolds B. W. and Vickers D. (1979) SR-0980-10, Oregon State University, Corvallis.

30. Eggleton A. E. J. and Thompson N. (1961) Nature 192, 935-6.

31. Sehmel G. A. (1983) Precipitation Scavenging, Dry Deposition and Resuspension (Pruppacher, Semonin and Slinn, eds.) Elsevier, 1013-1025.

32. Cawse P. A. (1974) AERE-R 7669, HMSO London.

33. Clough W. J. (1973) Aerosol Sci. 4, 227-234.

34. Dorran J. C. and Droppo J. G. (1983) Precipitation Scavenging, Dry Deposition and Resuspension. (Pruppacher, Semonin and Slinn eds.) Elsevier. 1003-1012.

35. Davies T. D. and Nicholson K. W. (1982) Deposition of Atmospheric Pollutants. Georgii and Pangrath (eds) Reidel.

36. Wesely M. L., Cook D. R., Hart R. C., Hicks B. B., Durham J. L., Speer R. E., Stedman D. H. and Tropp R. J. Precipitation Scavenging, Dry Deposition and Resuspension (Pruppacher, Semonin and Slinn, eds.) Elsevier, 943-952.

37. Wesely M. L., Hicks B. B., Dunnevik W. P., Frisella S. and Husar R. B. (1977) Atmos. Environ. 11, 561-563.

38. Katen P. C. and Hubbe J. M. (1983) Precipitation Scavenging, Dry Deposition and Resuspension (Pruppacher, Semonin and Slinn, eds) Elsevier, 953-962.

39. Sievering H. (1983) Precipitation Scavenging, Dry Deposition and Resuspension (Pruppacher, Semonin and Slinn, eds) Elsevier. 963-978.

40. Chadwick R. C. and Chamberlain A. C. (1970) Atmospheric Evironment, 4, 57-78.

41. Heinemann K., Vogt K. J. and Angeletti L. (1974) Atmosphere Surface Exchange of Particulate and Gaseous Pollutants, CONF 740921, 136-152.

42. Facchetti S. and Geiss F. Isotopic Lead Experiment - Status Report, EUR 8352 EN, CEC Joint Research Centre, Ispra.

43. Stewart K. (1967) Surface Contamination. B. R. Fish (ed) Pergamon. 63-74.

44. Sehmel G. A. (1983) Precipitation Scavenging, Dry Deposition and Resuspension (Pruppacher, Semonin and Slinn, eds) Elsevier, 1145-1159.

45. Sehmel G. A. (1983) Precipitation Scavengining, Dry Deposition and Resuspension (Pruppacher, Semonin and Slinn, eds) Elsevier, 1073-1086.

46. Garland J. A. (1979) AERE-R 9452, HMSO London.

47. Garland J. A. (1982) AERE-R 10106, HMSO London.

48. Sehmel G. A. (1973) Atmos. Environ. 7, 291-301.

49. Hodgin C. R. (1983) Precipitation Scavenging, Dry Deposition and Resuspension (Pruppacher, Semonin and Slinn, eds) Elsevier, 1175-1184.

50. Milham R. L., Schubert J. F., Watts J. R., Boni A. L. and Corey J. C. (1975) DP-MS-75-29. DuPont de Nemours & Co.

51. Garland J. A. (1983) Precipitation Scavenging Dry Deposition and Resuspension (Pruppacher, Semonin and Slinn, eds) Elsevier, 1087-1097.

52. Peirson D. H., Cawse P. A. and Cambray R. S. (1974) Nature 251, 675-679.

53. Pattenden N. J., Cambray R. S. and Playford K. (1981) Geochimica et Cosmochimica Acta. 45, 93-100.

54. Piotrowicz S. R., Duce R. A., Fasching J. L. and Weisel C. P. (1979) Mar. Chem. 7, 304-324.

55. Pattenden N. J., Cambray R. S. and Eakins J. D. (1983) Special Publication No. 3 of the British Ecological Society. Blackwell Scientific Publications, 259-271.

56. Eakins J. D., Lally A. E., Burton P. J., Kilworth D. R. and Pratley F. A. (1982) AERE-R 10127, HMSO London.

57.   Clough W. S. (1975) Atmos. Environ. $\underline{9}$, 1113-1119.

58.   Little P. and Wiffen R. D. (1977) Atmos. Environ. $\underline{11}$, 437-447.

TABLE I.   SOME RESULTS FROM THE CEC ISOTOPIC LEAD
           EXPERIMENT, TURIN, 1974-1983[a]

Sample Media	Isotopic Ratio $^{206}Pb/^{207}Pb$		
	Sample 1[b]	Sample 2[c]	Sample 3[d]
Petrol	1.186	1.060	–
Airborne particulate lead	1.174	1.064	–
Total deposition	1.155	1.130	36
Soil			
(0-5 cm)	1.156	–	–
(40-60 cm)	1.169	–	–
Leaves (Horsechestnut)	–	1.068	96
Nut (Horsechestnut)	–	1.145	21
Blood (adults)	1.163	1.134	32
(9-10 y)	–	1.127	39

[a]From Ref. 42.

[b]Before introduction of petrol of controlled isotopic composition.

[c]During period of complete substitution.

[d]Apparent percentage of lead of local airborne origin.

TABLE II. ARTIFICIAL TRACERS USED IN DEPOSITION AND RESUSPENSION
STUDIES

Tracer	Properties, Method of Application, Etc.
Radioactive tracers:	
^{35}S	Used to study SO_2 and sulphate deposition. Liquid scintillation counting.
^{131}I	Fission product, ? spectrometry.
^{59}Fe	Activation product, ? spectrometry.
^{212}Pb	Solid decay product of gaseous precursor, used as supersaturated vapor or to label powders. Short lived: ? spectrometry.
^{123}I, ^{64}Cu, ^{99m}Tc	Short-lived isotopes of low emission energy leading to low radiotoxicity. ? spectrometry.
Stable isotopes:	
Dy, In, Eu	Rare trace elements, neutron activation.

REVIEW OF PARTICLE TRACERS OF ATMOSPHERIC PROCESSES

Donald F. Gatz

Atmospheric Chemistry Section
Illinois State Water Survey
Champaign, IL 61820

ABSTRACT

This review is limited to particle tracers deliberately released for use in investigations of atmospheric processes, including transport, dispersion, wet and dry deposition, resuspension,
and others. Particle tracers have been used in atmospheric research since the early 1950s. Their use is currently declining in
favor of gaseous tracers useful over longer distances. Publications reporting applications to transport and dispersion peaked in
the late 1960s, those giving results of precipitation scavenging
and wet deposition experiments peaked in the early 1970s and
early 1980s. Many different materials have been used as tracers.
The materials generally fall into one of four groups: pollen and
spores, organic materials, inorganic materials, and radioisotopes.
This review describes the individual materials used as particle
tracers and also the methods used to generate an aerosol of them,
to sample them, and to analyze for them. A bibliography of over
100 papers is included.

I. INTRODUCTION

This paper reviews particulate materials deliberately released for use
as tracers of atmospheric processes. This includes releases of natural
materials, such as pollen, where the placement or timing of the release has
been manipulated by the experimenter. Other papers in this workshop cover
gaseous tracers released deliberately and both gas and particle tracers of
opportunity.

Because some of the keynote papers focus on tracer materials and
others focus on atmospheric processes for which tracer techniques may be
appropriate tools for investigation, this paper may overlap to some extent the
papers that address the use of tracers in investigations of atmospheric
processes.

Several major reviews of tracer applications have been published.
Dumbauld[1] provided an early review of fluorescent particle techniques for
measuring atmospheric dispersion. Islitzer and Slade[2] extensively reviewed
results of transport and dispersion experiments up to the mid-1960s. Gatz[3]
reviewed tracer experiments on precipitation systems. Sehmel's review[4] of

dry deposition includes results of many tracer experiments. Very recently, Johnson[5] reviewed tracer techniques for investigation of atmospheric transport and dispersion. A paper on tracer theory with applications in geosciences has also appeared.[6]

As indicated by the subjects of these reviews, particle tracers have been used to investigate atmospheric transport and dispersion,[2,7-37] precipitation scavenging and convective storm circulations,[38-70] and dry deposition.[9-13,17,71-89] Other stated objectives of atmospheric tracer experiments include investigations of stratospheric circulation and mixing,[90-92] atmospheric residence times of particles,[93] cooling tower mist interception by plants,[94] evaluation of cloud seeding effectiveness,[61,62,64,95,96] and particle resuspension[84,97-102] from the earth's surface, roads, or the leaves of plants.

Figure 1 shows a distribution of the number of published papers and reports on particle tracers located in my literature review, by year of publication (in 5-year blocks), for some major tracer applications. It appears that publications on particle tracers used in transport and dispersion experiments peaked in the late 1960s; those used in precipitation scavenging, wet deposition, and storm circulation experiments peaked in the early 1970s; and those used in dry deposition and resuspension experiments are at, or just past, their peak. Overall, use of particle tracers in atmospheric research appear to be declining in recent years.

II. MATERIALS USED AS PARTICLE TRACERS

A wide variety of materials have been used as particle tracers. Johnson[5] listed the key requirements for an effective atmospheric tracer: (1) Background concentrations of the tracer material in the atmosphere from natural and artificial sources must be small. (2) The tracer must follow air motions faithfully (or, if the tracer is simulating particles, as in a dry depositionexperiment, rather than an air parcel, its motions must be those of the particles it is simulating). (3) Transformations and/or removal (deposition) must be small, or at least well known and predictable (unless used in a deposition experiment). (4) The tracer must be easily handled and dispersed, at measureable rates.

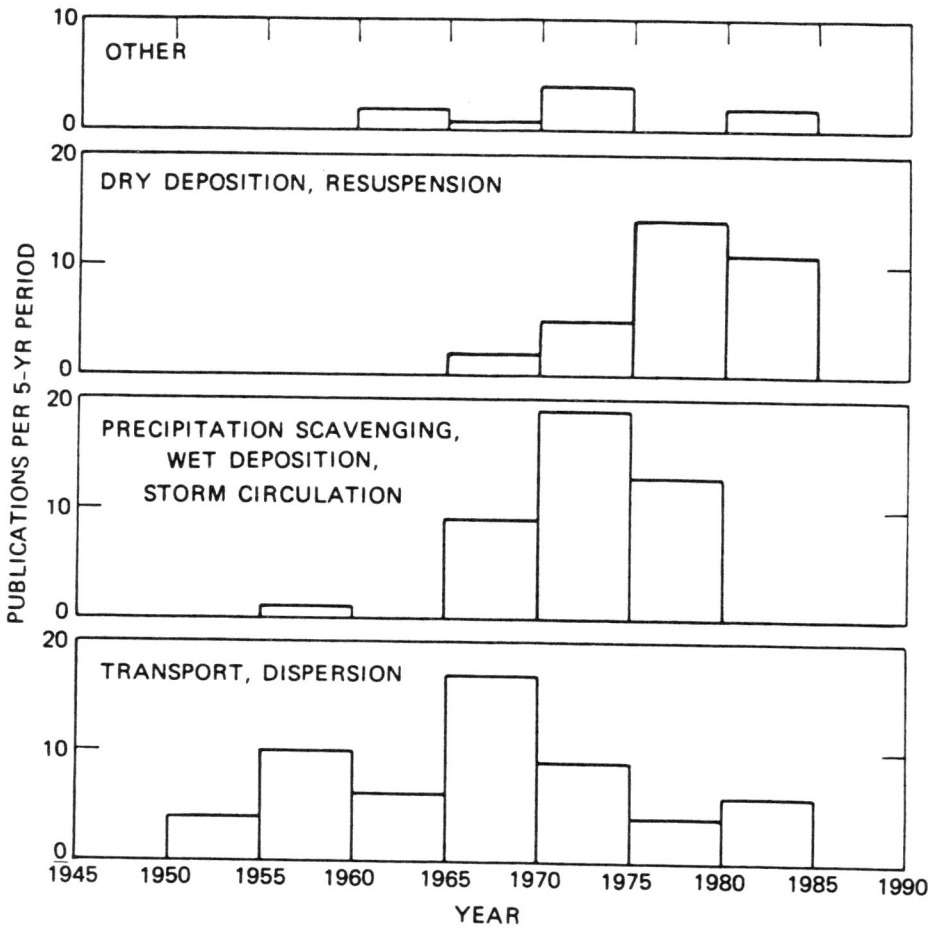

Fig. 1.   Time-frequency distribution of publication date (in 5-yr blocks) of papers and reports located in this literature search, by major application category.

(5) Sensitive analytical techniques must be available to enable very low concentrations of the tracer to be measured. (6) The tracer must be nontoxic and free from other adverse environmental effects. (7) The cost of the tracer and its sampling and analysis methods must be reasonable (at least, reasonable in terms of the benefits to be gained from the experiment).

For purposes of discussion, it seems reasonable to group the known tracer materials into four general categories: pollen and spores, organic materials, inorganic materials, and radionuclides. The following four subsections of the paper are each devoted to one of these categories. Information on individual tracer materials, generation methods, particle sizes, sampling methods, and analysis methods are presented primarily in tables.

A. Pollen and Spores

Table I lists 10 species of pollen and spores that have been used as atmospheric tracers. The table also gives information on the shape of the pollen grain or spore, as well as its surface characteristics and size. Many of those listed are spheroidal, including two that may collapse, but other shapes are represented as well. The surfaces range from smooth to spiny, and the diameters are from 1 µm to 18 x 58 µm. One advantage of pollen and spores as tracer materials is the rather uniform size of the grains within a given species. Pollen and spores have also been stained various colors and tagged with radioactivity to facilitate the identification of particles of the same size released at different times or different locations.

Methods for generating an aerosol of pollen or spores and for sampling and analyzing the grains are listed in Table II. A variety of methods have been used. Atomization methods usually involve suspending the particles in a liquid and spraying them through a nozzle or jet or using a spinning disk[103] to atomize the liquid into droplets containing, at the most, a single grain. Other methods involve dispensing single dry grains by jets of air or vibration of the dispenser. Pollen have also been released naturally from plants grown in a specific location, or in pots moved to the test site, or from plants treated to pollenate before their usual season. Bacillus subtilis spores were also sprayed on desert soil and mobilized by vehicle traffic or by marching or crawling men.

TABLE I. CHARACTERISTICS OF POLLEN AND SPORES[a]

Type	Shape	Surface	Diameter (μm)
Bacillus subtilis			1
Paper Mulberry (Broussonetra)	Spheroidal (may collapse)	Smooth	14
Ragweed (Ambrosia)	Spheroidal	Spiny	18-20
Timothy (Phleum)	Spheroidal (may collapse)	Smooth	30-35
Summer Cypress (Kochia scoparia)	Spheroidal	Smooth	30
Club Moss (Lycopodium)	Modified pyramidal	Slightly rough	32-33
Fern (Osmunda)	Spheroidal	Smooth	54
Fungus (Cronartium)	Pyriform acuminate	Smooth	18 ? 58
Castor Bean (Ricinus communis)	Ellipsoidal	Smooth	24 ? 38
Fern (Dryopteris)	Ovoidal	Slightly rough	33 ? 45

[a]Modified from Ref. 10.

TABLE II. GENERATION, SAMPLING, AND ANALYSIS METHODS
FOR POLLEN AND SPORES

A.   Bacillus subtilis

Generation:	Pneumatic atomizer Spinning disk generator Truck traffic (from soil) Men marching and crawling (soil)

Sampling:

Concentration:	Filters with preimpinger Impactor
Deposition:	Velvet paper Beaker
Analysis:	Counting by microscope

B.   All other pollen and spores.

Generation:	Pneumatic or atomizing nozzle Dry aerosol generator Air jet Aspirated, vibrated, dispensing bottle Natural, from specially grown or potted plants Natural, from plants treated to pollenate early

Sampling:

Concentration:	Slide-edge-cylinder Rotoslide sampler Sticky cylinders Wet and dry moss bags, grass, trays
Deposition:	Greased microscope slide
Analysis:	Counting by microscope

Sampling of airborne concentrations (Table II) has been accomplished with filters of various kinds, impactors, by wet and dry moss bags, grass, and trays, and by a number of devices that rely on wind impaction of the grains onto a sticky surface. Sampling of any large particles, including pollen, is difficult because their mass is large enough to keep them from closely following air motions, thus frequently resulting in anisokinetic sampling. The Rotoslide sampler, in which the pollen are sampled on the edge of a microscope slide rotated at high speed, is one way to overcome the wind speed effect. Even so, collection efficiencies generally do not approach 1.0.

Sampling pollen and spore deposition has usually been done using greased microscope slides,[11-14] although B. subtilis scavenged by snow flakes have also been sampled at the bottom of a chamber by ordinary laboratory beakers and by velvet paper.[42]

B. Organic Materials

Organic materials used as atmospheric tracers include one of the earliest--oil fog--as well as a number of fluorescent dyes and a chemiluminescent material. Information on these materials and their generation, sampling, and analysis methods is given in Table III.

Again, a wide variety of aerosol generation methods have been used. Smoke generators have been used to produce an oil fog that has droplets with diameters of 0.3 μm. Other organic tracer aerosols have been produced using various pneumatic spray nozzles or nebulizers to produce liquid droplets, which quickly evaporate and leave as residue the desired particles. With soluble tracers such as uranine dye, the concentration of the tracer in the sprayed liquid can be varied to control the size of the resulting particles. In general, higher concentrations produce larger particles.

Sampling methods often reflect the objectives of the experiment. The use of plants or precipitation collectors as sampling media shows experimental objectives related to dry and wet deposition, respectively. Methods for "sampling" oil fog include both qualitative and quantitative photographic techniques, as well as filter sampling and subsequent analysis by fluorescence measurements. In addition, oil fog provides a direct visual impression of transport and diffusion at a given site, which can be extremely valuable to a researcher, especially early in the investigation of an unfamiliar

TABLE III. TRACER MATERIALS SUMMARY -- ORGANIC MATERIALS

Material	Generation method	Particle diameter, μm	Sampling method	Analysis method[a]	Remarks[b]
Fluorescein[c]	Pneumatic nozzle		Filter	Fluorescence	D. L. = 2x10[-10]g
Glycerol[d]			String bean plants	Counting of Na-22 R A used to tag particles	
Oil fog[e] or smoke	Smoke generator	0.3	Photography Photometric densitometer Asbestos filter Nephelometer Lidar	Photography Densitometer Opacity method Fluorecence	
Oleic acid[f]	LaMer generator	0.5	Wet and dry moss bags, grass, and trays	Counting of R A. used to tag particles	
Polystyrene[g] spheres	Alcohol spray	3	Wet and dry moss bags, grass, and trays	Counting of R A. used to tag particles	
Rhodamine[g]	Jet atomization	0.4-6 0.8 MMD	Filter Bulk snow by shovel Precipitation samplers	Fluorescence	D.L. = 10[-10] g/mL Lee (Ref. 67) recommends against Rhodamine for scavenging experiments because of high natural background.
TMAE[h]	Bomblets	0.1-4	Film and video photography	Photography	
Uranine dye[i]	Nebulizer Collision generator Pneumatic nozzle Spinning disk Dry dust dispenser Modified commercial paint sprayer (rotating bell) Fluid atomizing generator Vibrating orifice	0.7 <1 2-10 (5.4 MMD) >2 7.5 17	Filter (isokinetic) Membrane filter Glass fiber filter Precipitation sampler Plant leaves (sunflower, tulip, poplar) Lidar	Fluorescence	Hygroscopic D L. = 10[-11] g/mL

[a]R.A. = radioactivity.
[b]D.L. = detection limit.
[c]Reference 22
[d]Reference 81.
[e]References 5, 16, 24, 25, 31.
[f]Reference 78.
[g]Reference 47, 55, 66, 67, 77
[h]Reference 7, 21.
[i]References 5, 23, 26, 47, 72-74, 79, 86.

location. TMAE, a chemiluminescent tracer, would appear to be capable of providing the same visual impression in nighttime situations.

Analysis methods for the organic materials were mostly fluorescence measurements. There were also some measurements of radioactivity, where tagged organic particles were used.

C. Inorganic Materials

This group, listed in Table IV, includes over 15 elements used as tracers. Often several different compounds of the same element have been used in the same, or associated experiments, where the different compounds were chosen for their contrasting chemical behavior (soluble vs insoluble) or surface characteristics (wettable vs nonwettable).

Aerosol generation methods include spray nozzles, nebulizers, and dry particle dispensers, as described previously, to dispense ZnS and ZnCdS fluorescent particles. The list of methods in Table IV also includes propane and acetone burners and pyrotechnic flares, all of which are also used to generate cloud-seeding materials. Some early experiments in the USSR used exploding artillery shells to inject tracer materials into convective storms.

The propane and acetone burners and the pyrotechnic flares often produce tracer particles with diameters near, or less than, 1 μm. The fluorescent particles are typically near 2 μm, but smaller and larger particles of this type have also been produced. In a case where a Na_2CrO_4 tracer was used to simulate cooling tower drift, a much larger size, typical of mist droplets, was produced by a spray nozzle.

A wide range of sampling methods has been used for the inorganic tracer materials, again reflecting diverse experimental objectives, including wet and dry deposition, resuspension, and dispersion measurements. Sampling media include precipitation samplers, filters, impactors, and also a real-time sampler for Li particles that uses flame ionization measurements.

Several analytical methods have been employed, typically those used for metal analyses, including NAA, AAS, ASV, XRF, and ICAP (acronyms defined in Table IV).

D. Radioisotope Materials

Table V lists several radioisotopes used as tracers. Others were listed on earlier tables when they were used for the purpose of tagging the primary

TABLE IV. TRACER MATERIALS SUMMARY -- INORGANIC MATERIALS

Material	Generation method	Particle diameter, μm	Sampling method	Analysis method[a]	Remarks
$AgNO_3$[b]	Propane AgI burner		Precipitation samplers	NAA	D.L. 10^{-11} molar
AgI[c]	Propane burner Acetone burner		Precipitation sampler network	NAA	
Ag[d]	Propane AgI burner Pyrotechnic flare Droppable pyrotechnic flare	0.05-0.1 0.03 (mean)	Snow collector Sequential precipitation collector, mobile collector	NAA	
CuS[e]	Rocket shell explosion		Precipitation samplers		
$CuSO_4$[f]		3-7	Paper, metal, soil, plants		
$CaMoO_4$[g]	Wind erosion		Cascade impactor	XRF (Mo)	
Cs[h]	Pyrotechnic flare		Precipitation sampler netowrk	AAS	
$DsNo_3$, $Sc(NO_3)_3$ $Sr(NO_3)_2$[i]	Propane AgI burner		Precipitation samplers	NAA	
FeOOH (beta)[j]	Atomizer		Filter sampler array	R.A. counting (^{59}FE)	
Fe_2O_3[k]	Wind erosion	2 and 5	Filters	R.A. counting	
$Fe(OH)_3$[k]	Wind erosion		Filters	R.A. counting	
In_2O_3[l]	Acetone burner	0.05-0.35			Nonhygroscopic
In[m]	Pyrotechnic flare	<1	Precipitation collectors	NAA AAS	Nonwettable
LiCl[n]	Acetone burner	0.27-0.63, mean = 0.45	Mobile precipitation collector	ASV	Soluble
Li_2CO_3[o]	Sonic atomizing nozzle Modified commercial paint sprayer (rotating bell) Spray nozzle	0.5-1.5 f1.6 0.7	Flame ionization detector (real time) Filters	Flame ionization ICAP	Low solubility
Li stearate[p]	Acetone burner Mechanical dispenser		Precipitation sampler	NAA	Nonwettable
Li[q]	Acetone burner pyrotechnic flare	0.4	Precipitation sampler network	NAA AAS	Soluble
Na_2CrO_4[r]	Pneumatic nozzle	100-1300, 55% between 500-800	2-yr grass, pine, poplar plants, filter disks	Radioactivity counting (Cr-51)	

TABLE IV. TRACER MATERIALS SUMMARY -- INORGANIC MATERIALS (concluded)

Material	Generation method	Particle diameter, μm	Sampling method	Analysis method[a]	Remarks
$(NH_3)_2SO_4$[s]	Ultrasonic nebulizer	0.7	Conc: filter, impactor. Depos: filter, funnels, bucket	Liquid scintillation counting (S35)	
Ir[t]	Acetone burner		Precipitation sampler network	NAA	Nonwettable
Ta, Re, Au, Eu, Ru, Os[u]	Acetone burner		Precipitation sampler network	NAA	
ZnS, SnCdS[v]	Pneumatic atomizer Smoke generator Dry particle dispenser (Unspecified) Air jet Exploding cannister Vehicle traffic Modified commercial paint sprayer Dry duster Venturi tube	1-2 2.3 2.7 1-10 (mean 3.2) 5 <10 <25	Rotorod sampler Filters Real-time sampler Velvet paper, beaker (snow crystals) Impactor Wet and dry moss bags, grass, trays Drum impactor	Manual counting (by microscope) Real-time sampler Radioactivity counting (tagged particles)	D.L. 2×10^{-10}g

[a]AAS = atomic absorption spectrophotometry; ASV = anodic stripping voltammetry; ICAP = inductively coupled argon plasma; NAA = neutron activation analysis; XRF = X-ray fluorescence.
[b]Reference 41.
[c]References 55, 58, 59, 66.
[d]References 61, 62, 64, 70, 93, 95, 104.
[e]Reference 63.
[f]Reference 105.
[g]Reference 99.
[h]Reference 54.
[i]Reference 41.
[j]Reference 89.
[k]Reference 84.
[l]Reference 106.
[m]References 40, 46, 49, 55, 57, 58, 64-66.
[n]Reference 106.
[o]References 17, 73-75, 85.
[p]References 55, 66.
[q]References 53, 55, 58-60, 65, 66.
[r]Reference 94.
[s]Reference 71.
[t]Reference 66.
[u]References 43-45, 65.
[v]References 8, 18, 20, 22, 27, 28, 30, 32, 42, 74, 77, 78, 83, 92, 100, 107-112.

TABLE V. TRACER MATERIALS SUMMARY -- RADIOISOTOPES

Material	Generation method	Particle diameter, μm	Sampling method	Analysis method[a]	Remarks
Au-198[b]		0.78 AMAD[c]	String bean plants	R.A. counting	
Cu-64[d]	Pneumatic nozzle	9.5	Conc: Rotoslide, Depos: greased microscope slides Airborne gamma counter	R.A. counting	
P-32[e]	Rocket shell explosion		Precipitation sampling netowkr	R.A. counting	Hygroscopic
Te-131[f]	Rocket shell explosion		Precipitation sampler network	R.A. counting	Nonhygroscopic
Po-210[g]	Rocket shell explosion		Precipitation sampler network	R.A. counting	
Pu-238[h]	Pneumatic nebulizer		Bean plants, membrane filters, cascade impactors, thermal precipitators	R.A. counting	
Rh-102[i]	Nuclear explosion		Filter	R.A. counting	
W-185[i]	Nuclear explosion		Filter	R.A. counting	

[a]R.A. = radioactivity.
[b]Reference 81.
[c]AMAD = activity mean aerodynamic diameter.
[d]References 113-115.
[e]References 51, 52.
[f]Reference 51.
[h]Reference 82.
[i]Reference 91.

tracer particle to enable more convenient analysis (such as radioactive tagging of pollen or fluorescent particles that would otherwise have to be counted manually).

We can divide the list of Table V into three separate groups. The first group includes radioisotopes produced specifically as tracers for particles injected into the stratosphere by nuclear test explosions (Rh-102 and W-185). The second group includes several nuclides injected into precipitation systems by artillery shells (P-32, Te-131, and Po-210). The third group is composed of two radionuclides used in dry deposition experiments (Cu-64, Au-198) and a cloud circulation experiment in Australia (Cu-64).

Sampling again reflects the purpose of the various experiments, and analyses were by radioactivity counting methods.

III. CONCLUDING REMARKS

Over 100 papers on tracer experiments in atmospheric research were reviewed. A wide variety of tracer materials have been applied to such atmospheric processes as transport, dispersion, wet and dry deposition, resuspension, and others. This paper gives a general description of the tracers and their applications. For more details on specific applications, please consult the original papers, listed in the References section.

Particle tracers are frequently used as one componenet of dual-tracer experiments. Two or more different tracers with different particle sizes may be used together to measure deposition relative to one another. Or, a particle tracer may be used with an inert gas tracer. In this case, the inert gas is used to correct the particle concentrations for the effects of turbulent diffusion, so that particle deposition may be estimated.

Particle tracers are ordinarily useful only to distances of a few 10s of kilometers. Interest is now turning to transport and dispersion over much greater distances of 100 to 1000 km. Thus, the use of particle tracers appears to be declining, although they are still being employed for short-range transport and dispersion, dry deposition, and resuspension studies.

Oil fog, although one of the earliest tracer materials, can still provide a clear visual impression of transport and dispersion at an unfamiliar site, and is thus extremely useful in guiding further experiments.

REFERENCES

1. Dumbauld, R. K., 1962: Meteorological tracer technique for atmospheric diffusion studies. J. Appl. Meteorol., 1 (4): 437-443.

2. Islitzer, N. F. and D. H. Slade, 1968: Diffusion and transport experiments. Chapter 4, In: D. H. Slade, Meteorology and Atomic Energy-1968. U.S. Atomic Energy Commission, Division of Technical Information.

3. Gatz, D. F., 1977: A review of chemical tracer experiments on precipitation systems. Atmos. Environ., 11: 945-953.

4. Sehmel, G. A., 1980: Particle and gas dry deposition: a review. Atmos. Environ., 14: 983-1012.

5. Johnson, W. B., 1983: Meteorological tracer techniques for parameterizing atmospheric dispersion. Journal of Climate and Meteorology, 22: 931-946.

6. Nir, A. and B. L. Kirk, 1982: Tracer theory with applications in geosciences, Part 1, Oak Ridge National Lab report ORNL-5695/P1, Environ. Sci. Div., Publication No. 1964, 93 pp.

7. Elrick, R. M. and R. E. Smith, 1967: Investigation of TMAE as a chemiluminescent atmospheric tracer. Sandia Corp., Albuquerque, NM, 53 pp.

8. Leighton, P. A., W. A. Perkins, S. W. Grinnell, and F. X. Webster, 1965: The fluorescent particle atmospheric tracer. J. Appl. Meteorol., 4: 334-348.

9. Raynor, G. S., J. V. Hayes, and E. C. Ogden, 1974: Particulate dispersion into and within a forest. Boundary-Layer Meteorol., 7: 429-456.

10. Raynor, G. S., J. V. Hayes, and E. C. Ogden, 1975: Particulate dispersion from sources within a forest. Boundary-Layer Meteorol., 9 (3): 257-277.

11. Raynor, G. S., E. C. Ogden, and J. V. Hayes, 1970: Dispersion and deposition of ragweed pollen from experimental sources. J. Appl. Meteorol., 9 (6): 885-895.

12.   Raynor, G. S., E. C. Ogden, and J. V. Hayes, 1972: Dispersion and deposition of timothy pollen from experimental sources. Agricultural Meteorology, 9, (1971/1972): 347-366.

13.   Raynor, G. S., E. C. Ogden, and J. V. Hayes, 1972: Dispersion and deposition of corn pollen from experimental sources. Agronomy Journal, 64: July-August, 1972, 420-427.

14.   Raynor, G. S., E. C. Ogden, and J. V. Hayes, 1973: Dispersion of pollens from low-level, crosswind line sources. Agricultural Meteorology, 9, (1971/1972): 347-366.

15.   Raynor, G. S., E. C. Ogden, and J. V. Hayes, 1976: Dispersion of fern spores into and within a forest. Rhodora, 78, 815: 473-487.

16.   Raynor, G. S., P. Michael, R. M. Brown, and S. Sethu Raman, 1975: Studies of atmospheric diffusion from a nearshore oceanic site. J. Appl. Meteorol., 14 (6): 1080-1094.

17.   Sehmel, G. A., 1981: A dual-tracer experiment to investigate pollutant transport, dispersion, and particle dry deposition at the Rio Blanco oil shale site in Colorado, Preprint from Second Conf. on Mountain Meteorology, Nov. 9-12, 1981, Steamboat Springs, CO, pp 137-146.

18.   Hilst, G. R. and N. E. Bowne, 1971: Diffusion of aerosols released upwind of an urban complex. Environ. Sci. Technol., 5 (4): 327-332.

19.   Webster, F. X., 1963: Collection efficiency of the Rotorod FP Sampler. Tech. Report No. 98, Aerosol Lab., Metronics Assoc., Inc., Stanford Industrial Park, Palo Alto, CA.

20.   Dotson, W. L., P. W. Nickola, and M. A. Wolf, 1968: Real time sampling of zinc sulfide tracer in diffusion studies. In: Mawson, C. A., Ed., Proceedings, U.S. Atomic Energy Commission Meteorological Information Meeting, Sept. 11-14, 1967, Chalk River, Ontario, Canada, 277-291.

21.   Smith, R. E. and R. M. Elrick, 1968: Investigation of TMAE (tetrakis dimethylamino ethylene), a chemiluminescent material, as an atmospheric tracer. In: C. A. Mawson, Ed., Proceedings, U.S. Atomic Energy Commission Meteorological Information Meeting, Sept. 11-14, 1967, Chalk River, Ontario, Canada, 334-346.

22. Ludwick, J. D., 1966: Atmospheric diffusion studies with fluorescein and zinc sulfide particles as dual tracers. J. Geophys. Res., 71 (6): 1553-1558.

23. Gussman, R. A. and A. M. Sacco, 1964: Notes on a method of establishing a size distribution of meteorological tracer aerosols. J. Appl. Meteorol., 3 (5): 638-640.

24. Smith, M. E., et al., 1958: The variations of effluent concentrations during temperature inversions. J. Air Poll. Control Assoc., 7: 194-197.

25. Barad, M. L. and B. Shorr, 1954: Field studies in diffusion of aerosols. Amer. Ind. Hyg. Assn. Quart. 15: 136-140.

26. Robinson, E., J. A. McLeod, and C. E. Lapple, 1959: A meteorological technique using uranine dye. J. Meteorol., 16: 63-67.

27. Crozier, W. D. and B. K. Seely, 1955: Concentration distributions in aerosol plumes three to twenty-two miles from a point source. Trans. Amer. Geophys. Union, 36, 42-52.

28. Wedin, B., N. Frossling, and B. Aurivillius, 1959: Comparison of concentration measurements of sulphur dioxide and fluorescent pigment. Advances in Geophysics, 6: New York, Academic Press, 425-427.

29. Hay, J. S. and F. Pasquill, 1957: Diffusion experiments from a fixed source at a height of a few hundred feet in the atmosphere. J. Fluid Mech., 2.

30. Braham, R. R., B. K. Seely, and W. D. Crozier, 1952: A technique for tagging and tracing air parcels. Trans. Amer. Geophys. Union, 33: 825-833.

31. Gifford, F. A., 1980: Smoke as a quantitative atmospheric diffusion tracer. Atmos. Environ., 14: 1119-1121.

32. Bierly, E. W. and G. C. Gill, 1963: A technique for measuring atmospheric diffusion. J. Appl. Meteorol., 2: 145-150.

33. McElroy, J. L., 1969: A comparative study of urban and rural dispersions. J. Appl. Meteorol., 8: 19-31.

34. Goodman, J. K. and A. Miller, 1977: Mass transport across a temperature inversion. J. Geophys. Res., 82: 3463-3471.

35. Fritschen, L. K., R. Edmonds, 1976: Dispersion of fluroescent particles into and within a Douglas fir forest. Atmos-Surf. Exc., Partic. Gaseous Pollut. (1974), ERDA Symposium Series 38, R. J. Engelmann and G. A. Sehmel, Coordinators, Proc. of Conf., Richland, WA, pp 280-301.

36. Islitzer, N. F., 1961: Short-range atmospheric-dispersion measurements from an elevated source. J. Meteorol., 18 (4): 443-450.

37. Raynor, G. S., L. A. Cohen, J. V. Hayes, and E. C. Ogden, 1966: Dyed pollens and other spores as tracers in dispersion and deposition studies. J. Appl. Meteorol., 5: 728-729.

38. May, F. G., 1958: The washout of Lycopodium spores by rain. Quart. J. Roy. Meteorol. Soc., 84: 451.

39. Dingle, A. N. and K. S. Bhatki, 1970: Tracer indium determination in rain samples by neutron activation and radiochemical analysis. Radiochem. Radioanal. Letters, 3 (1): 71-79.

40. Bhatki, K. S. and A. N. Dingle, 1969: The measurement of tracer indium in rain samples. J. Appl. Meteorol., 9 (2): 276-282.

41. Haller, W. A. 1967: Evaluation of elemental tracers and neutron activation analyses in atmospheric precipitation studies. In: Pacific Northwest Laboratory Annual Report for 1966 to the USAEC Div. of Biology & Medicine, Vol. II, Physical Sciences. Part I, Atmospheric Sciences, BNWL-481, pp 82-84.

42. Sood, S. K. and M. R. Jackson, 1969: Scavenging study of snow and ice crystals. Final Report No. IITRI-C6105-9 of IIT Research Institute Technology Center.

43. Young, J. A., T. M. Tanner, C. W. Thomas, and N. A. Wogman, 1975: The entrainment of tracers near the sides of convective clouds. Pacific Northwest Laboratory Annual Report for 1974 to the USAEC Division of Biomedical and Environmental Research, BNWL-1950 PT3, UC-11, pp 140-142.

44. Young, J. A., T. M. Tanner, C. W. Thomas, and N. A. Wogman, 1976: The entrainment of tracers near the sides of convective clouds. BWL-2000, PT3, Pacific Northwest Laboratory Annual Report for 1975. Battelle, Pacific Northwest Laboratory, Richland, WA, pp 179-184.

45. Young, J. A., T. M. Tanner, C. W. Thomas, and N. A. Wogman, 1976: The entrainment of tracers into convective clouds at 10,000 to 13,500 feet near St. Louis. Proceedings of Precipitation Scavenging Symposium (1974), Champaign, IL, Oct. 14-18, 1974. U.S. Energy Research and Development Administration.

46. Davis, W. E., 1976: An estimate of in-cloud scavenging of a tracer injected into two frontal storms. BNWL-2000, PT3, Pacific Northwest Laboratory Annual Report for 1975, Battelle, Pacific Northwest Laboratory, Richland, WA, pp 160-165.

47. Dana, M. T., 1970: Scavenging of soluble dye particles by rain. In: Engelmann, R. J., and W. G. N. Slinn, Coords., Precipitation Scavenging (1970), Proc. Sympos., Richland, WA, June, 1970. U.S. Atomic Energy Commission, Division of Technical Information, pp 137-147.

48. Engelmann, R. J., R. W. Perkins, D. I. Hagen, and W. A. Haller, 1966: Washout coefficients for selected gases and particles. Report BNWL-SA-657, Battelle Pacific Northwest Laboratory, Richland, WA.

49. Dingle, A. N., D. F. Gatz, and J. W. Winchester, 1969: A pilot experiment using indium as tracer in a convective storm. J. Appl. Meteorol., 8 (2): 236-240.

50. Dana, M. T. and J. M. Hales, 1976: Statistical aspects of the washout of polydisperse aerosols. Atmos. Environ., 10: 45-50.

51. Burtseva, I. E., L. V. Burtseva, and S. G. Malakhov, 1970: Washout characteristics of a ^{32}P aerosol injected into a cloud. In: Atmospheric Scavenging of Radioisotopes, Symposium Proceedings, Palanga, USSR, June 7-9, 1966, pp 242-250. Published for Env. Sci. Serv. Admin., US Dept. of Commerce and NSF by Israel Program for Sci. Translation, Jerusalem.

52. Shopauskas, K., B. Styra, E. Verba, B. K. Verbrene, et al., 1970: Washout of radioisotopes injected into a cloud from data of ground observations. In: Atmospheric Scavenging of Radioisotopes, Symposium Proceedings, Palanga, USSR, June 7-9, 1966, pp 233-241. Published for Env. Sci. Serv. Admin., US Dept. of Commerce and NSF by Israel Program for Sci. Translation, Jerusalem.

53. Gatz, D. F., 1974: METROMEX: air and rain chemistry analyses. Bull. Amer. Meteor. Soc., 55: 92-93.

54. Staff, Illinois State Water Survey, 1975: Projects of the Illinois State Water Survey, In: Auer, A. H., Compiler, 1975 Operational Report for METROMEX, Univeristy of Wyoming, Laramie, WY, pp 13-18.

55. Slinn, W. G. N., 1973: In-cloud scavenging studies. Annual Report to the U.S. Atomic Energy Commission, Division of Biomedical and Environmental Research, Vol. II, Part 1, pp 76-79, Battelle Pacific Northwest Laboratory, BNWL-1751 PT1.

56. Thomas, C. W., R. G. Rieck, J. A. Young, and N. A. Wogman, 1973: An aerosol generator system for sequential tracer releases. Pacific Northwest Laboratory Annual Report for 1972. Battelle Pacific Northwest Laboratory, Richland, WA. BNWL-1751, PT1, pp 101-103.

57. Dingle, A. N., 1976: Scavenging and dispersal of tracer by a self-propagating convective shower system. Proceedings, Precipitation Scavenging Symposium (1974), Champaign, IL, Oct. 14-18, 1974. U.S. Energy Research and Development Admin.

58. Atkinson, D. G., W. E. Davis, L. F. Radke, B. C. Scott, et al., 1972: Precipitation scavenging of tracers released into frontal storms. Annual Report to U.S. Atomic Energy Commission, Division of Biology and Medicine, Battelle Pacific Northwest Laboratory, BNWL-1651, PT1, pp 1-16.

59. Smith, M. D., D. H. Steffe, and T. J. Henderson, 1969: Chadron tracer experiment. Report 69-2 to National Science Foundation, Chadron Atmospheric Research Institute, Chadron State College, Chadron, Neb.

60. Semonin, R. G., 1972: Tracer chemical experiments in midwest convective clouds. Symposium Proceedings, Rapid City, SD, June 26-29, 1972, pp 83-87, Third Conf. on Weather Modification, Amer. Meteorol. Soc., Boston.

61. Summers, P. W., 1972: The silver fallout patterns in precipitation from seeded convective storms. Symposium Proceedings, Rapid City, SD, June 26-29, 1972, pp 279-286, Third Conf. on Weather Modification, Amer. Meteorol. Soc., Boston.

62.   Renick, J. H., A. J. Chisholm, and P. W. Summers, 1972: The seedability of multicell and supercell hailstorms using droppable pyrotechnic flares. Symposium Proceedings, Rapid City, SD, June 26-29, 1972, pp 272-278, Third Conf. on Weather Modification, Amer. Meteorol. Soc., Boston.

63.   Preobrazhenskaia, E. V., 1968: Detection of copper traces in precipitation in cases of cuprous sulfide modification of convective clouds. Glavnia Geofisicheskaia Observatoriia, Trudy, No. 224: 169-175.

64.   Warburton, J. A., 1973: The distribution of silver in precipitation from two seeded Alberta hailstorms. J. Appl. Meteor., 12: 677-682.

65.   Davis, W. E., J. A. Young, and J. M. Thorp, 1974: Progress report on in-cloud scavenging in frontal storms. Annual Report to U.S. Atomic Energy Commission, Division of Biomedical and Environmental Research, Part 3, pp 109-111 Battelle Pacific Northwest Laboratory, BNWL-1850 PT3.

66.   Davis, W. E. and J. A. Young, 1976: Results of in-cloud tracer releases in frontal storms. Proceedings of Precipitation Scavenging Symposium, 1974, Champaign, IL, Oct. 14-18, 1974. U.S. Energy Research and Development Administration.

67.   Lee, R. N., 1975: A reevaluation of the use of Rhodamine for in-cloud scavenging studies. Pacific Northwest Laboratory Annual Report for 1974. Battelle Pacific Northwest Laboratory, Richland, WA, BNWL-1950 PT3, pp 134 -136.

68.   Engelmann, R. J., 1963: Rain Scavenging of particulates. General Electric Co., Richland, WA, Hanford Atomic Products Operation, 227 pp.

69.   Engelmann, R. J., 1963: Rain Scavenging of zinc sulphide particles. J. Atmos. Sci., 22: 719-724.

70.   Engelmann, R. J., D. I. Hagen, W. A. Haller, and R. W. Perkins, 1966: Washout coefficients for silver iodide. Report, BNWL-SA-798, Battelle Pacific Northwest Laboratory, Richland, WA.

71.   Ibrahim, M., L. A. Barrie, and F. Fanaki, 1983: An experimental and theoretical investigation of the dry deposition of particles to snow, pine trees and artificial collectors. Atmos. Environ., 17, No. 4: 781-788.

72.   Sehmel, G. A. and S. L. Sutter, 1974: Particle deposition rates on a water surface as a function of particle diameter and air velocity. J. Rech. Atmos., 911-919.

73.   Sehmel, G. A. and W. H. Hodgson, 1984: Generation of nearly monodispersed particles for dry deposition field experiments. PNL Annual Report for 1983 to the DOE Office of Energy Research, Part 3, Atmospheric Sciences, pp 41-42.

74.   Sehmel, G. A., 1984: Dry deposition experiments using multiple tracers on the Hanford diffusion grid. Pacific Northwest Laboratory Annual Report for 1983. PNL-5000 Pt3. Battelle, Pacific Northwest Laboratory, Richland, WA.

75.   Sehmel, G. A., 1982: Plume depletion by particle dry deposition in a forested canyon. Pacific Northwest Laboratory Annual Report for 1981. PNL-4100 Pt3. Battelle, Pacific Northwest Laboratory, Richland, WA.

76.   Sehmel, G. A. and W. H. Hodgson, 1975: Penetration of depositing particles through crushed gravel. Pacific Northwest Laboratory Annual Report for 1974, BNWL-1950, Part 3, Battelle Pacific Northwest Laboratory, Richland, WA, 181-183.

77.   Nickola, P. W. and R. N. Lee, 1975: Direct measurements of particulate deposition on a snow surface. Pacific Northwest Laboratory Annual Report for 1974, BNWL-1950, Part 3, Battelle Pacific Northwest Laboratory, Richland, WA pp 181-183.

78.   Clough, W. S., 1975: The deposition of particles on moss and grass surfaces. Atmos. Environ., 9: 1113-1119.

79.   Sehmel, G. A. and W. H. Hodgson, 1976: Field deposition velocity measurements to a sagebrush canopy. Pacific Northwest Laboratory Annual Report for 1975: Atmos. Sciences, BNWL-2000-3. Battelle, Pacific Northwest Laboratory, Richland, WA, 89-91.

80.   Raynor, G. S., 1976: Experimental studies of pollen deposition to vegetated surfaces. Atmos-Surf. Exch., Partic. Gaseous Pollut. (1974). ERDA Symposium Series 38, R. J. Engelmann and G. A. Sehmel, Coordinators, Proc. of Conf., Richland, WA. pp 264-279.

81.   Vaughn, B. E., 1976: Suspended particle interactions and uptake in terrestrial plants. Atmos.-Surf. Exch. Partic. Gaseous Pollut. (1974). ERDA Symposium Series 38, R. J. Engelmann and G. A. Sehmel, Coordinators, Proc. of Conf., Richland, WA. pp 228-243.

82.   Craig, D. K., B. L. Klepper, and R. L. Buschbom, 1976: Deposition of various plutonium-compount aerosols on to plant foilage at very low wind velocities. Atmos.-Surf. Exch., Partic. Gaseous Pollut. (1974). ERDA Symposium Series 38, R. J. Engelmann and G. A. Sehmel, Coordinators, Proc. of Conf., Richland, WA. pp 244-263.

83.   Nickola, P. W. and G. H. Clark, 1976: Field measurement of particulate plume depletion by comparison with an inert gas plume. Atmos.-Surf. Exch., Partic. Gaseous Pollut. (1974). ERDA Symposium Series 38, R. J. Engelmann and G. A. Sehmel, Coordinators, Proc. of Conf., Richland, WA. pp 74-86.

84.   Garland, J. A., 1983: Some recent studies of the resuspension of deposited material from soil and grass. Precipitation Scavenging, Dry Deposition, Resuspension, Elsevier, New York, pp 1087-1097.

85.   Sehmel, G. A., 1983: Particle dry deposition measurements with dual tracers in field experiments. Precipitation Scavenging, D. D., Resus., pp 1013-1025.

86.   Wedding, J. B., R. W. Carlson, J. J. Stukel, and F. A. Bazzazz, 1975: Aerosol deposition on plant leaves. Environ. Sci. Technol., 9: 151-153.

87.   Chamberlain, A. C., 1960: Aspects of the deposition of radioactive and other gases and particles. Int. J. Air Pollut., 3: 63-88.

88.   Chamberlain, A. C., 1966: Transport of Lycopodium spores and other small particles to rough surfaces. Proc. R. Soc., 296, 45-70.

89.   Garland, J. A., 1983: Dry deposition of small particles to grass in field conditions. Precipitation Scavenging, Dry Deposition, Resuspension, Elsevier, New York, pp 849-858.

90.   Feely, H. W. and J. Spar, 1960: Tungsten-185 from nuclear bomb tests as a tracer for stratospheric meteorology. Nature, 188 (4756): 1062-1064.

91.   Kalkstein, M. I., 1962: Rhodium 102 high-altitude tracer experiment. Science, 137: 645.

92.   Droessler, E. G., K. J. Heffernan, and E. K. Bigg, 1967: A stratospheric air tracer experiment using zinc sulfide. J. Appl. Meteorol., 6: 373-379.

93.   LaCaux, J. P. and J. A. Warburton, 1983: Precipitation scavenging of submicron particles released by rockets into convective storms. Precipitation Scavenging, Dry Deposition, and Resuspension, Elsevier, New York, pp 303-313.

94.   Taylor, F. G., Jr., P. D. Parr, and F. Ball, 1980: Interception and retention of simulated cooling tower drift (100 1300 μm diameter) by vegetation. Atmos. Environ., 14: 19-25.

95.   Parungo, F. P, and C. E. Robertson, 1969: Silver analysis of seeded snow by atomic absorption spectrophotometry. J. of Appl. Meteorol., 8 (3): 315-321.

96.   de Pena, R. G., and E. A. Caimi, 1967: Hygroscopicity and chemical composition of silver iodide smoke used in cloud seeding experiments. J. Appl. Meteorol., 24: 383-386.

97.   Aylor, D. E., 1976: Resuspension of particles from plant surfaces by wind. Atmos.-Surf. Exch., Partic. Gaseous Pollut. (1974). ERDA Symposium Series 38, R. J. Engelmann and G. A. Sehmel, Coordinators, Proc. of Conf., Richland, WA. pp 791-812.

98.   Hereim, A. T. and B. Richie, 1976: Resuspended bacteria from desert soil. Atmos.-Surf. Exch., Partic. Gaseous Pollut. (1974). ERDA Symposium Series 38, R. J. Engelmann and G. A. Sehmel, Coordinators, Proc. of Conf. Richland, WA. pp 835-845.

99.   Sehmel, G. A., 1983: Resuspension rates from aged inert tracer sources. In: H. R. Pruppacher, R. G. Semonin, and W. G. N. Slinn, Eds., Precipitation Scavenging, Dry Deposition, and Resuspension, Elsevier, New York, pp 1073-1086.

100.   Sehmel, G. A., 1976: Particle resuspension from an asphalt road caused by car and truck traffic. Atmos.-Surf. Exch., Partic. Gaseous Pollut. (1974). ERDA Symposium Series 38, R. J. Engelmann and G. A. Sehmel, Coordinators, Proc. of Conf., Richland, WA. pp 859-882.

101.   Sehmel, G. A., F. D. Lloyd, 1974: Particle resuspension rates Atmos.-Surf. Exch., Partic. Gaseous Pollut. (1974). ERDA Symposium Series 38, R. J. Engelmann and G. A. Sehmel, Coordinators, Proc. of Conf., Richland, WA. pp 859-882.

102.   Sehmel, G. A., 1973: Particle resuspension from an asphalt road caused by car and truck traffic. Atmos. Environ., 7 (3): 291-209.

103. Harstad, J. B., M. E. Filler, W. T. Hushen, and H. M. Decker,
     1970: Homogeneous bacterial aerosols produced with a spinning-disc
     generator. Appl. Microbiol. 20, (1): 94-97.

104. Warburton, J. A., L. G. Young, 1972: Determination of silver in
     precipitation down to 10-11 m concentrations by ion exchange and
     neutron activation analysis. Anal. Chem., 44 (12): 2043-2045.

105. Herbert, M., K. J. Vost, L. Angeletti, 1976: Studies on the deposition
     of aerosols on vegetation and other surfaces. Report, ORNL tr-
     4314, Oak Ridge National Lab. Oak Ridge, TN.

106. Finnegan, W. G., L. A. Burkhardt, and P. St. Amand, 1971:
     Generation of atmospheric tracer materials. Proc., Int. Conf.
     Weather Modification, Canberra, Australia. Australian Academy of
     Science and American Meteorological Society Boston, MA. 355 pp.

107. Nickola, P. W., 1964: Sampler for recording the concentration of
     airborne zinc sulfide on a real time scale. U.S. Atomic Energy
     Comm., BNWL-36, Sec. 1, Battelle Pacific NW Lab., Richland, WA.
     pp 59-64.

108. Nickola, P. W., 1966: Instrumentation of the Queen Air Aircraft for
     sampling of zinc sulfide atmospheric tracer. In: PNL annual report
     for 1965, Vol. I: (Atmos. Sci.) Battelle Pacific Northwest Lab, p 11.

109. Orgill, M. M. and P. W. Nickola, 1974: Evaluation of an airborne
     fluorescent particle counter for atmospheric tracer studies. Pacific
     NW Lab Annual Rept. for 1973, Part 3. Atmospheric Sciences.
     Battelle Memorial Inst., Richland, WA. pp 75-79.

110. Perkins, W. A., P. A. Leighton, S. W. Grinnell, F. X. Webster,
     1952: A fluorescent atmospheric tracer technique for
     mesometeorological research. Proc. Nat'l. Air Poll. Symp., 2nd,
     Pasadena, CA pp 42-46.

111. Perkins, W. A., and R. W. McMullern, 1962: The change in particle
     size distribution of fluorescent particle tracer during city wide
     travel. Metronics Assoc., Inc. Palo Alto, CA, Aerosol Lab., Contract
     CWB 10313, TR-95, 15 pp.

112. Murray, J. A. and L. M. Vaughan, 1970L Measuring pesticide drift at
     distances to four miles. J. Appl. Meteorol., 9 (1): pp 79-85.

113. Fookes, R. A., J. S., Watt, and J. A. Warburton, 1962: A radioisotope technique for tracing air movement in clear air and in clouds. Nature, 196: 328-329.

114. Raynor, G. S. and M. E. Smith, 1964: A diffusion-deposition tracer system. Brookhaven National Lab, Upton, NY, BNL 859 (T-343), 17 pp.

115. Rayno, G. S., 1964: Radioactive copper as an atmospheric tracer. Preprint, Am. Soc. of Agricultural Engr., Saint Joseph, MI, 33 pp. (Presented at Am. Soc. of Ag. Engrs., Winter Meeting, New Orleans, LA, Dec. 8-11, 1964, paper 64-823.)

ATMOSPHERIC TRACERS OF OPPORTUNITY FROM IMPORTANT CLASSES OF AIR POLLUTION SOURCES

Glen E. Gordon

Department of Chemistry
University of Maryland
College Park, MD 20742

ABSTRACT

Atmospheric tracers unique to various classes of air pollution sources are needed for study of the transport, dispersion, transformation, and deposition of materials released by these sources, especially as they contribute to atmospheric problems of major concern such as acid deposition and visibility degradation. By far the greatest amount of work on these "tracers of opportunity" has involved the use of elemental concentrations on airborne particles. Indeed, on an urban scale, particles from a number of sources can be identified and their contributions calculated by receptor-modeling techniques. These sources and some key elements associated with them include coal-fired plants (As, Se), oil combustion (V, Ni), motor vehicles (Pb, Br), refuse incineration (K, Zn, Sb, In, Cd), concrete and cement (Mg, K, Ca), and sulfide ore smelters (Cu, In, Se). Despite the success of receptor models based on elements, other tracer methods should and are being developed, especially for sources that emit mainly carbonaceous materials, e.g., vehicles burning diesel fuel and nonleaded gasoline; home heating by oil, gas, or wood; and refineries and petrochemical plants. Organic compounds could provide useful signatures, but unknown amounts of them may react with other species before reaching receptor sites. Isotopes of some elements will be quite useful, especially ^{14}C for determining the fraction of "modern" vs fossil carbon. Permanent gases and vapor-phase species of moderately volatile elements should be considered. Considerable progress is being made on the use of morphology and compositions of individual particles. Tracers are also needed for studies on a regional and global basis, but for these longer distance scales, serious questions must be answered regarding fractionation and modification of species by deposition and reactions during transit.

I. INTRODUCTION

By the term "tracers of opportunity," I refer to materials that are released to the atmosphere by certain types of sources in the normal course of their operations and which can be used to trace the movement of those materials after release. Ideally, such a tracer should be unique to that source; i.e., none of it is released by any other significant kind of source. In practice, this ideal situation almost never occurs, but there are cases in which it is almost true. For example, in most cities, especially near the

North Atlantic Coast of the U.S., most of the V and Ni result from combustion of oil. The more usual situation is that, although a particular source may dominate the emissions of a species, corrections must be made for the amounts contributed by other sources. Below, I show how the amounts of each species contributed by each of several types of sources can be determined simultaneously by a matrix technique called "chemical mass balances" (CMBs).

Although emissions involving the energy industries or major energy consumers (e.g., tranportation) are of primary concern to DOE for the reasons noted above, we must consider all types of sources simultaneously. Most work done to date in this field has involved the observation of trace elements borne by particles, so this review necessarily focuses on them. However, other kinds of tracers are under development and the potential for their applications is also covered.

II. CHEMICAL MASS BALANCES

According to the CMB approach,[1] the particle-borne atmospheric concentration of element i, C_x, during a particular sampling period can be expressed as the sum of contributions for each type of source, j:

$$C_i = \sum_j m_j x_{ij}, \tag{1}$$

where m_j = mass concentration of airborne particulate matter
     contributed by source j

and x_{ij} = is the concentration of element i in particulate matter
     from source j.

To apply Eq. (1), we must know the concentrations of the elements of interest in particles from all the important sources in the area, i.e., the x_{ij} values. Ideally, one should collect and analyze particles from all the important sources in the area in question, but as costs of these measurements are often beyond the resources available for a study, source composition terms taken from the growing libraries of data may suffice.[2] Given the ambient concentrations, C_i, and the source composition matrix, x_{ij}, the objective is to find the source strength terms, the $\underline{m}_j$ values, that best fit the observed concentrations. This is normally done by a least squares fitting procedure involving matrix inversion.[3]

Results for a few of the 40 elements measured in the Washington D.C. area atmosphere and fitted by CMBs are listed in Table I.[4] Although 28 elements were used in the least-squares fit, the elements shown are some of the most important for determining the contributions of particles from the assumed seven sources: soil, sea salt, limestone, motor vehicle emissions and combustion of coal, oil and refuse. As can be seen in Table I, the contributions from sea salt, limestone, motor vehicles, and combustion of oil and refuse can be determined from the concentrations of Na, Ca, Pb, V, and Zn after corrections for small contributions from other components are made. An historic problem for CMBs has been the resolution of soil and coal contributions.   Coal contains small fragments of aluminosilicate material that are released when the coal is burned. Pollution controls usually collect 90% or more of these particles as precipitator or scrubber ash, but the composition pattern of those released, at least for major elements such as Al, Si, and Fe, is quite similar to that of soil. Fortunately, some of the trace elements are quite useful for resolving contributions for these sources. Arsenic is a chalocophillic element and, thus, enriched in materials such as coal, which contain substantial amounts of S. Therefore, As (and some other chalcophiles) is a good marker for emissions from coal combustion in many urban areas. Manganese is also helpful for resolving coal and soil, as it is depleted coal relative to soil.

III. TRACERS FOR SEVERAL TYPES OF SOURCES

In Table II, I have used the results of the Washington CMBs and other data to determine the elements that arise in major amounts from a single source. As discussed above, As appears to originate mainly from coal combustion. Selenium is a very interesting case. As it is chemically quite similar to S, one would expect coal to be a major source. The figure of 90% indicates that 90% of the predicted Se, based on the CMBs, arises from coal combustion. However, the total Se predicted from the seven sources is considerably less than the observed particulate Se. Thus, the Se from coal-fired plants accounts for only 31% of the observed Se. Originally, we had thought that the discrepancy arose because the Se concentration in the coal component was based on the analyses of particles collected directly from the hot stack gases,[4] whereas it is known that the majority of Se is in the vapor phase at stack temperatures,[5] but it apparently partially condenses at ambient temperatures.[6] This is indeed part of the problem, but recent studies on the regional sulfate component in the rural area of Shenandoah Valley, VA, demonstrate that a significant portion of the Se not accounted for in Washington is brought in from long distances away.[7] Thus, whereas Se appears to be a good tracer for coal combustion, it will include both local and distant sources. In a recent study of sources in the Philadelphia area, Howes et al.[8] sampled a coal-fired power plant by dilution source sampling, which appears to have raised the Se concentration relative to other elements several-fold. If we make this change in the component used for Washington CMBs and include the regional sulfate component, it should be easy to account for the observed Se. Perhaps surprisingly, particulate I appears to arise in significant amounts from coal combustion, its other major source being sea salt. Until it is studied in more detail, however, particulate I is somewhat dangerous to use because we know little about its gas/particle ratio and how the ratio varies in time and space.

Clearly V and Ni are excellent tracers for emissions from oil-fired plants. The CMBs of Washington[4] severely underpredicted Ni apparently because the component used, based on emissions from a single plant, had an unusually low ratio of $Ni/V(0.17)$. If a component based on the recent studies of Philadelphia sources were used, in which the Ni/V ratios are close to unity, oil combustion would fully account for Ni in the Washington atmosphere.

Lead and bromine arise almost completely from motor vehicles in Washington and most other cities. It is safer to use Pb as the tracer, as some of the Br forms volatile species that leave the particles.[9,10] Although leaded gasoline is gradually being phased out, particulate emissions from vehicles burning nonleaded gasoline are so low that they are insignificant relative to those burning leaded gasoline. Unfortunately, there is no reliable elemental tracer of diesel emissions. An estimate of the particulate load contributed by diesel trucks can be made on the basis of Pierson and Brachaczek's Allegheny Mt tunnel studies if one knows the relative amounts of diesel and light-vehicle traffic in the area.

In the Washington area, refuse emission is a major source of Ag, Zn, and Sb and contributor of appreciable amounts of several other elements, including Cd, In, and Cu.

To achieve a reasonable fit to alkaline earths, especially Ca, in most cities, it is usually necessary to include limestone or some chemically similar component. The origin of this material has not been clearly established, but one can imagine many possibilities, including gravel dust, quarries and cement plants, construction and demolition of buildings, erosion of streets and buildings, and agricultural liming.

Virtually every element is present in airborne soil; however, most other sources contribute some of the prominent elements of soil, especially emissions from coal-fired power plants, as discussed above, so it's not easy to find a good soil tracer. Because of the deficiency of Mn in coal, Mn is a fairly good tracer for soil, especially among coarse particles. However, there has long been an unidentified source of fine Mn. Recent studies in Shenandoah Valley[7] and Watertown, MA[11] suggest that much of the unexplained Mn comes into eastern cities as a part of the regional sulfate component. Of course, Mn and other alloying elements such as Cr, V, Ni, and Co must be treated with great care in areas of steelmaking. Despite its large contribution to particulate loading, the composition pattern of entrained soil is not well established. A variety of studies suggest that the fine soil particles that become entrained have a composition different from that of bulk soil. It is important that the fractionation of soil be better understood. Even if one is not concerned with the large amount of total suspended particulate matter (TSP) contributed by soil, because soil contributes appreciable amounts of many elements, these contributions must be determined accurately to obtain reliable residuals to be fitted by other components. As there is no single tracer element that comes entirely from soil, it is probably best to determine its contribution and those of other components by least-squares fitting to a large suite of elements, e.g., 28 as in Washington.[4]

Another source of increasing importance, especially since the energy crises of the 1970s, is wood smoke. In many areas it has become a very large source of TSP and fine carbon particles, and a significant cause of visibility degradation. Recently, Lewis and Einfeld[12] conducted a study in Albuquerque, NM, with results that are summarized in Table II. During the daytime, wood burning accounted for 45% of the K, 51% of volatile carbon and 13% of nonvolatile (or "elemental") carbon borne by fine particles (diameter ? 2.5 μm). Most of the remainder of the particulate carbon, plus virtually all of the Pb and Br arose from motor vehicles. At night, the portions of fine K, volatile carbon, and nonvolatile carbon arising from wood burning increase to an estimated 95, 88, and 88%, respectively. Three important points should be made in connection with these results. First, one can obtain much more definitive information by separating the particles into two or more size groups than by using data from filter collections. Particles of diameter 2.5 μm are mostly made up of soil, limestone, and the fly-ash portion of emissions from coal-fired power plants. The fine fraction usually contains much more information because of strong enrichments of chalcophile and other moderately volatile elements. If the Albuquerque samples had included coarse particles, a large correction for K from soil would have been required, increasing errors in the remainder coming from wood burning. Second, if there are any refuse incinerators in the area, their atmospheric contributions must be carefully determined because they are a strong source of fine K. Third, to calculate the TSP contributed by wood smoke from the K concentration, one must know the types of wood being burned, as the K content of the particles varies by an order of magnitude among different woods.

Regarding carbonaceous sources, a very important, unambiguous tracer is coming into use, namely, the ratio of ^{14}C to ^{12}C (or total carbon). Carbon-14 is produced continuously by cosmic-ray interactions and is maintained at an approximately steady-state ratio to stable ^{12}C and ^{13}C in the atmosphere. Living material incorporates carbon having approximately the steady-state ratio, but when life stops, the incorporated ^{14}C decays with its characteristic 5,700-yr half life. Measurement of the $^{14}C/^{12}C$ ratio is the basis of radiocarbon dating of old artifacts that contain carbon.

The same principle is used to determine the portions of recently living carbonaceous material, i.e., "modern" carbon, and of fossil carbon, as the latter has been out of steady-state with atmospheric carbon for so many half lives that it contains virtually no ^{14}C. In the classic use of this method, Cooper et al.[14] used this technique to determine that up to 44% of carbon borne by particles in suburban Portland, OR, was modern carbon, apparently most of it the result of wood burning. Until the past decade, ^{14}C was assayed by observing its ß-decay in very low backgrond proportional counters, but even with the best current technology, this required mg quantities of carbon. However, the development of nuclear accelerator methods has reduced the sample requirement to 10's of µg.[15] With this great increase in sensitivity, it has now become practical to measure the $^{14}C/^{12}C$ ratio of individual compounds separated from atmospheric samples and, thus, determine fossil vs. modern contributions to them. The ratio of $^{13}C/^{12}C$ also varies among different sources and can be used to provide an added dimension of information, although the results are not as strikingly unambiguous as for $^{14}C/^{12}C$.

The comments made above regarding tracers for sources apply to areas that have only the kinds of sources that are common to most areas. The picture may be quite different in areas that have large special sources such as steel mills or sulfide ore smelters. The latter, for example, usually emit large amounts of many chalcophile elements. In areas where they are located, they could well be the dominant sources of As and Se rather than coal-fired power plants, and of Ag, Zn, and Sb rather then incinerators. One particular element that is a strong tracer for smelters is In, which has been used to observe emissions from the Sudbury, Canada, area as far away as Rhode Island. Parrimton et al.[16] have observed emissions for the smelter region of Arizona at Mauna Loa Observatory in Hawaii.

IV. REGIONAL SCALE TRACERS

Most of the above discussion has involved urban-scale tracers of primary particles from mostly local sources. Efforts are under way to develop tracers that would be useful on a regional scale. They are needed to help us understand the long-range transport of species involved in acid deposition and visibility degradation. Rahn and Lowenthal,[17] for example, have devised a tracer system based on the ratios of As, In, Sb, Zn, and noncrustal V and Mn concentrations to that of Se, where noncrustal means that one subtracts contributions from entrained soil. They have identified characteristic patterns of these ratios for particles originating in six large regions of the world, including three in North America: Southern Ontario, Canada (SONT), the east coast of the U.S. (ECOAST), and the interior of the U.S. (INT), which is basically the area from about the Mississippi River eastward to Pennsylvania.

Ratios for the three North American regions based on Rahn and Lowenthal's work are listed in Table III. Although there are other small differences between the regions, the most distinctive features are the high V/Se ratios for the ECOAST pattern (because of oil-burning in that region) and the high In/Se ratio for SONT because of the smelters there. The INT pattern has a higher As/Se ratio than of that the ECOAST, probably because of extensive coal burning in the INT region.

Our group at Maryland has begun to investigate the compositions of particles from various areas, which are brought to rural sites by long-range transport. Our initial study[7] involved instrumental neutron activation analysis (INAA) of the fine fractions of 33 samples of particles collected in the Shenandoah Valley by Stevens et al. and previously analyzed by x-ray fluorescence (XRF).[18] Back-trajectories were calculated for each sampling period by using the Air Resources Laboratory (ARL) Branching Atmospheric Trajectory (BAT) model[19] and the samples grouped according to wind direction.

Ratios of concentrations for particles borne by winds from the west and southeast are compared with those of Rahn and Lowenthal in Table III, except for In/Se, as In levels were below detection on a majority of the Shenandoah Valley samples. We see many of the same trends as Rahn and Lowenthal, e.g., high V/Se ratios in the east, although there are considerable differences of average ratios. We are not too concerned about differences for the ECOAST group, as ours were based on air masses coming from the Norfolk area, whereas theirs refer to the Washington-Boston corridor. Of greater concern are the differences for the INT pattern. Their samples were collected in Underhill, VT, when the winds were from the Ohio River Valley region and may well have been modified during transit through PA and NY.

Although Rahn and Lowenthal's tracer system represents an interesting first step in the development of regional scale tracers, their approach has some potential weaknesses. Some of the patterns are based on very few samples and will probably change as more data are obtained. Their ratio patterns are based on air-filter samples, which include both coarse and fine particles. As noted above, the fine particles carry most of the information and are the ones transported over long distances, so the inclusion of coarse particles from mostly local sources confuses the issue. Furthermore, I question the validity of the idea that a huge region such as the interior U.S.

has a single pattern of particle compositions. There are surely considerable differences between particles from the Ohio River Valley and those from Chicago, St. Louis, or Detroit. In order to investigate this question, we are now applying INAA to samples collected at three rural sites in the Ohio River Valley, which were previously analyzed by XRF.[20] Back trajectories were used to select samples collected from air masses that were transported from various areas of the midwes'. The data on additional trace elements (obtainable by INAA) of these selected samples should clarify this issue. It would be most useful for the future of regional scale modeling if we could find unique elemental patterns associated with various subregions of the Midwest.

One implicit assumption in the regional scale modeling done to date, certainly in the work of Rahn and Lowenthal,[17] is that the particles bearing the tracer elements do not become fractionated during transit. In our work,[7] we have exclusively used data from the ? 2.5-μm-diameter particle fractions from samplers. Although Rahn and Lowenthal have used air-filter data, five of their seven elements are borne almost entirely by fine particles; by sub-tracting the crustal contributions of Mn and V, they obtain residuals that should mostly be present on fine particles.

To test the coherence of the fine-particle group, Tuncel et al.[7] per-formed factor analysis to identify the elements that accompany sulfate at the Shenandoah Valley site and multiple linear regressions to determine the relative amounts of species borne by the "regional" sulfate. The associated species are found to be H^+ and NH_4^+ ions, Se, Mn, As and very small amounts of Sb, Zn and In. The amounts of Se and Mn associated with regional sulfate are sufficient to account for significant amounts of these elements in the urban areas of Boston, MA, and Washington, D.C., accounting for some of the previously unexplained Se and fine Mn (see above).

Since most of the sulfate and Se in rural areas surely arise from coal-fired power plants, at least when the wind is from the west, the question arises as to whether or not the regional sulfate component simply represents fine particles released from coal-fired plants with the addition of SO_4^{2-}, H^+, and NH_4^+ during transit. By comparison of the relative concentrations of several elements in the regional sulfate component with those of fine par-ticles from coal-fired power plants, they found agreement within about a factor of two for Mn, Zn, Se, In, and Sb, but less As in the regional component than expected. Some of this discrepency may result from rather large variations of As/S ratios in various coals[21], but the results may also mean that there are significant additional sources of these elements.

One of the most surprising and important result is that the substantial amounts of Al and Fe present on 2.5-μm diam particles from coal combustion were not observed in the regional sulfate in Shenandoah Valley.[7] This means that there is fractionation of particles even when one selects the fine particle fraction. Ondov et at.[22] subdivided particles from a coal-fired plant into several fractions with diam 2.4 μm, observing considerable structure in this region. The finest fractions, with diam ~0.1 μm, had much greater ratios of enriched elements to lithophiles such as Al and Sc than the particles of about 2-μm diam. It is not clear if the fractionation of these particles during long-range transport occurs simply on the basis of size, which would be surprising, or if it has something to do with the physical chemical properties of the particles. The particles bearing SO_4^{2-}, Se, and other trace elements are much more soluble than those containing mainly lithophile elements, so these different groups of particles will be affected differently if processed by clouds and fogs during transit.

The structure and fractionation of fine particle groups must be investigated in connection with the development of regional scale tracers of opportunity. Samples were collected with a micro-orifice cascade impactor during the Deep Creek Lake experiment of 1983 and the samples are now being analyzed.[23] In very preliminary results, Ondov[23] has observed considerable structure among the very fine particles, with the particles bearing different elements centered about different diameters. If this result holds up as more data become available, emissions from various sources can perhaps be associated with different size groups as well as different compositions. Thus, studies of the very fine particles are not only necessary for establishing regional scale tracing, but they present an opportunity to develop another dimension for identification of sources.

V. OTHER POTENTIAL TRACERS

The above discussion has focused mostly on the use of particle-borne elements as tracers, but many other species are under consideration and development as noted in Table IV. One of the most obvious possibilities is organic compounds, of which hundreds of thousands are known, providing enormous opportunities for observation of unique patterns associated with different sources and areas. For many

Table I.   Chemical Mass Balances for Several Elements Borne by Particles in the Washington, D.C. Area (130 samples, Aug./Sept., 1976)[a]

Element	Predicted contribution (ng/m³)								Observed
	Soil	Limestone	Coal	Oil	Refuse	Motor Vehicle	Marine	Total	(Ng/m³)[b]
Na	43	0.8	8.3	12	35	-	201	300	300±20
Al	812	9	517	0.4	6	-	-	1340	1350±110
K	154	6	67	0.4	47	13	7	295	400±20
Ca	66	635	47	8.2	7	47	7.6	820	860±40
V	1.1	0.04	1.6	23	0.01	-	-	26	25±2
Mn	13	2.3	1.6	0.1	0.3	1.3	-	18	25±30
Fe	511	8.3	362	2.8	2.8	34	-	920	1000±60
Ni	0.46	0.04	1.0	4.0	0.07	0.34	-	6.0	17±2
An	1.1	0.04	2.6	1.6	51	7.3	-	64	85±6
As	0.06	0.002	3.1	0.03	0.10	-	-	3.32	3.25±0.20
Se	0.001	0.0002	0.78	0.035	0.016	0.035	-	0.87	2.5±0.2
Br	0.1	0.01	2.1	0.05	0.66	167	1.25	171	136±9
Ag	0.001	0.0001	-	0.006	0.23	-	-	0.24	0.20±0.01
Cd	0.001	0.0001	0.13	0.003	0.64	1.03	-	1.8	2.4±0.2
In (pg/m³)	0.7	0.1	2.3	<0.1	2.4	-	0.4	5.9	20±1
Sb	0.008	0.0004	0.13	0.007	0.89	0.60	-	1.6	2.1±0.2
I	0.058	0.003	2.1	-	-	-	1.14	3.3	2.0±0.1
Pb	0.15	0.02	2.1	0.39	34	428	-	465	440±20

[a]Ref. 4.

[b]Avg.±std. deviation of the mean value.

Table II. Airborne Contributions of Certain Marker Species

Source	Elements (% from source)
Washington, D.C. (all sizes)[a]	
Coal-fired plants	As (94), Se (31-90), I (64)
Oil-fired plants	V (88), Ni (100)
Motor vehicles	Br (98), Pb (92)
Refuse incineration	Ag (96), Zn (77), Sb (56)
Limestone/concrete	Ca (77), Mg (37)
Soil	Mn (72)
Albuquerque (fine particles)[b]	
Wood burning	K (45-95), Vol. C (51-88), Elem. C (13-88)
Mobile sources	Pb, Br (100), Vol. C (19-49), Elem. C (23-63)
Other	
Sulfide smelters[c]	In and other chalcophiles
Modern carbon[d]	14C

[a]Ref. 4.   [b]Ref. 12.   [c]Ref. 13.   [d]Ref. 14.

Table III. Geometric Mean Ratios of Several Trace Element Concentations (Fine Fraction) to Se Based on Work of Rahn and Lowenthal[a] and Tuncel et al.[b]

	Ratio and geometric std. deviation					
	$\frac{(Mn)}{Se}$ncr	$\frac{(V)}{Se}$ncr	$\frac{Zn}{Se}$	$\frac{As}{Se}$	$\frac{Sb}{Se}$	$\frac{In}{Se}$(x1000)
Rahn and Lowenthal signatures						
ECOAST	5.7 (1.6)	13.8 (1.7)	31 (1.3)	0.58 (2.9)	0.80 (1.8)	7.8 (1.6)
INT	2.6 (1.5)	1.96 (1.4)	10.8 (1.3)	0.92 (1.2)	0.28 (1.4)	3.9 (1.7)
SONT	13.9 (1.1)	1.77 (1.9)	57 (1.1)	8.0 (1.2)	0.75 (1.2)	46 (1.7)
Tuncel et al., Shenandoah Valley						
Trajectories from Westerly directions	1.49 (1.6)	0.60 (1.8)	6.6 (1.6)	0.24 (2.3)	0.20 (2.9)	-
Trajectories from southeast	1.33 (1.1)	6.1 (4.5)	9.0 (2.6)	0.49 (1.9)	0.23 (1.7)	-

[a]Ref. 17.

[b]Ref. 7.

Table IV. Other Potentially Useful Tracers

Particles

    Organic compounds - possible tranformations
    Isotopes - stable (e.g., C, S, B) and radioactive (heavy element)
    Individual particles - morphology, composition
    Mineral phases

Gases

    Permanent - SO_2, NO_x, light carbon species
    Vapor phase - organic compounds, halogens, Hg, Se, B
    Isotopes - (see above)

years, the study of organic compounds was hindered by the lack of sensitive analytical techniques and the consequent requirement for very large samples. Over the past decade, this situation has improved dramatically as a result of development and improvements in techniques such as gas and liquid-phase chromatography coupled with mass spectrometry. Although these methods are still not, in general, as sensitive as those for trace elements, it is now practical to analyze samples collected with high volume samplers for a wide range of compounds.[2,24]

The other major problem for the use of organic compounds is their possible transformation between the source and receptor. Reactive species such as OH, HO_2, NO_3, O_3, photons, etc., can transform some compounds into others during transit. Until the rates at which various organic compounds are destroyed can be determined, or it is established that certain classes of compounds are not appreciably destroyed during urban and/or regional transit, concentrations of organic compounds cannot be safely used in receptor models. Little work has been done in the field to investigate these problems;[25] however, they could be studied by measuring concentrations of organic species relative to that of a conservative tracer, either intentional or of opportunity, at the source and the receptor. Some of the tracers listed in Table II could be used for this purpose, e.g., K for wood burning, Pb for motor-vehicle emissions. If there are other significant sources of the organic compounds or the tracer, one must correct for the other contributions or, as Daisey has demonstrated, use multiple regressions of organic compounds vs. the tracer.[26]

For organic compounds that remain in the gas phase, one could turn the problem into an advantage: if rate constants for the major destructive path(es) are known, the average concentration of the destructive reactants could be determined from the amount of destruction. Concentrations of species that are very difficult to measure directly could be measured in this way. Fehsenfeld demonstrated the use of a similar approach to determine the OH radical concentration.[27] It is doubtful that the destruction of organic compounds attached to particles can be treated in such a systematic way.

Another potentially useful tracer system that is just beginning to be exploited is the observation of the morphology and composition of individual particles. The large particles can be observed with optical microscopes, which are capable of resolving characteristic particles from certain sources, e.g., cenospheres from coal-fired power plants.[28] Although large particles are useful for identifying local sources, fine particles are needed for regional studies. A major impediment to the study of individual particles, especially fine particles, has been the enormous amount of personnel time needed to observe enough particles to obtain good statistics for particles from minor sources. That problem is being overcome by the development of automatic, computer-assisted techniques for scanning electron microscopes (SEMs). The apparent two-dimensional size and shape of each particle, plus estimates of major-element concentrations (based on XRF) are compared with a stored library of these characteristics for particles from many types of sources.[28,29] As experience is gained with this method, it may become extremely valuable. In principle, it should be capable of resolving particles from several dozen sources, whereas there is some natural upper limit on the number of sources that can be resolved by trace-element concentrations (a function of the uncertainties of the analytical measurements and the variations of concentration patterns among different sources). Many promising new single-particle methods have been only slightly used including secondary ion mass spectrometry (SIMS) and laser microprobe mass analyses (LAMMA).[30]

Yet another identification method is x-ray diffraction (XRD) to detect the presence of various mineral phases.[31] It is useful to observe mineral phases associated with major sources such as soil, limestone, coal fly ash, and combustion of leaded gasoline, but it is not clear that it will be useful for very minor sources. The sample size required has been very large (~200 $\mu g/cm^2$), but the recent development of an XRD system specifically designed for air-filter samples promises to greatly reduce the sample needed.[32]

Some of the most critical questions about sources may be answerable by the measurement of permanent gases and other vapor-phase species. When we focus so much attention on trace species, it is easy to forget that some of the most difinitive information comes from the measurement of major pollutant gases from sources. Under many circumstances, virtually all of some gases originate from a particular class of source, e.g., SO_2 from coal-fired plants, CO from transportation. These gases have the advantage of being measurable in real time with response times of a few seconds, allowing one to follow rapid changes that cannot be observed with most other methods discussed above.

Trace inorganic gases and vapors may be quite useful for source identification. It is known, for example, that major portions of B, Hg, Se, As, and halogens leave the stacks of coal-fired power plants in the gas phase.[5,33-36] With the possible exception of As, major portions of these elements remain in the gas phase even after the stack emissions are cooled by mixing with ambient air.[35-39] Coal-fired plants are, of course, not the only sources of vapor-phase species. Much of the halogens released by motor vehicles and the oceans (as sea salt) end up in the gas phase. Other high temperature sources such as incinerators and oil-fired power plants are known or thought to be important sources of vapor-phase species,[34] so we cannot at present attribute these species to a certain class of sources. However, further study of trace gases might allow some of these species to be developed as valuable tracers.

Two vapor-phase elements may ultimately be of particular value for tracing emissions from coal-fired plants over long distances, namely, B and Se. Selenium is chemically so similar to sulfur that they will generally have the same sources. In many areas, most airborne sulfur arises from coal-fired power plants, so it seems safe to assume that most of the Se does, too (and there is experimental evidence supporting this assumption[7,18]). Similarly, much of the B in coal leaves the stack in the gas phase.[35,40] Coal-fired plants are certainly a major source of airborne B, but not enough is now known to rule out the importance of other possible sources (e.g., glass plants). Indeed, in marine areas, a considerable amount of gas-phase B comes from the oceans.[35,36]

If it can be established that in some areas of the U.S., e.g., non-marine areas such as the Midwest, virtually all of the airborne Se and B arise from coal combustion, these species can be used as valuable tools to observe the transformation and deposition of sulfur species from coal-fired plants. Lewis and Stevens[41] recently outlined the basis of a "hybrid" receptor model, which avoids the enormous uncertainties of the usual regional sulfate models by taking ratios of concentrations of sulfur species to those of other species released from coal-fired power plants. In this way, absolute errors in many poorly known quantities involving transport, dilution, and deposition cancel out of the hybrid model. Boron and Se would be particularly valuable in this application, because it appears that B remains in the gas phase[35,36] (as H_3BO_3), whereas, much of the Se becomes attached to particles fairly soon after release. If these assumptions can be established, one could trace the behavior of SO_2 gas via gas-phase B and that of sulfate particles by that of particulate Se. We are assuming that H_3BO_3 gas has the same diffusion and deposition properties as those of SO_2 and that Se-bearing particles behave the same as S after the latter is converted to particulate form. Many of these points are simply assertions at present, but could be tested by well desigend experiments.

Finally, other potentially valuable tracers are the abundances of isotopes (both stable and radioactive) for certain elements. The use of ^{14}C and ^{13}C is discussed above. Stable isotopes of several other light elements (giving large enough fractional mass differences to be differentiated by processes) may be quite useful, e.g., B and S. Some source materials, such as coal, also contain substantial amounts of natural radioactive heavy elements, which could serve as the basis of identifying their emissions.

VI. SUMMARY AND RECOMMENDATIONS

Many atmospheric problems are so difficult to study by absolute methods (i.e., regional sulfate and acid problems) that they can be handled much more reliably by conducting measurements relative to tracers of certain sources or areas. The tracers can be either intentionally released materials or tracers of opportunity, which are normally associated with the source or area. Intentional tracers have the advantage that they can be totally controlled: the time, location, and altitude of release can be chosen as well as the phase and chemical properties of the tracer.

Tracers of opportunity have complementary advantages relative to intentional tracers. For one thing, there is obviously no cost of the tracers themselves, which can become a major cost for tracing over distances of hundreds of km. One has little, if any, control of the release of the tracers. On the other hand, it is virtually impossible to release intentional tracers in such a way that they simulate various situations, e.g., all coal-fired power plants in the Midwest, area sources with dimensions of a city or state. Tracers of opportunity are, of course, released in the chemical and physical state typical of their sources and have the same behavior after release. Thus, both types of tracers are needed, depending on the kind of problem to be solved.

Most tracers of opportunity developed so far are elements borne by particles, but many other kinds of tracers are needed and under development, especially organic compounds, inorganic gases and vapors, individual particle characteristics, both stable and radioactive isotopes, and certain mineral phases.

Before the use of tracers of opportunity can be fully developed, many fundamental studies are needed, especially including the following:

- More complete measurements of the compositions and physical characteristics of materials released from a variety of important sources, conducted with dilution source or plume sampling to simulate changes that occur after materials mix with ambient air;

- Analysis of separate size fractions of particles with diam < 2.5 μm in both ambient air and source emissions;

- Analysis of samples of various types at rural sites in many areas, coupled with local meteorology as well as calculation of back-trajectories from National Weather Service data to establish concentration patterns associated with various regions and major sources;

- Vertical profiles for many species, at least up to cloud-forming levels so as to use tracers of opportunity to gain an understanding of in- and below-cloud processes;

- Gas/particle ratios for many moderately volatile inorganic tracers, especially B and Se, at sources, in plumes and in ambient air to test assumptions needed for use of these species in hybrid receptor models;

- Measurement of concentrations of organic compounds, both particulate and gas phase, at sources and receptors, relative to conservative tracers for those sources, to determine rates of destruction of the organic compounds under a range of atmospheric conditions; (note that a considerable amount of this work is planned as part of the Integrated Air Cancer Project of EPA, with an initial focus on motor vehicles and wood burning.[42])

- Further development of the use of $^{14}C/^{12}C$ and $^{13}C/^{12}C$ ratios for source identification by applying the measurements to specific compounds or classes of compounds;
- Investigation of the feasibility of using other isotopic abundances, both stable and radioactive, for receptor-modeling purposes;
- Development of the use of individual particle methods and XRD for mineral phases for receptor-model applications.

ACKNOWLEDGEMENTS

The ideas in this paper are based on many present and previous studies at the University of Maryland especially involving Ilhan Olmez, Semra Tuncel, Ann Sheffield, Josef Parrington, William Zoller, Gregory Kowalczyk, and Scott Rheingrover, plus collaborators from other institutions including Robert Stevens and Thomas Dzubay of EPA. Those works were supported by many grants and contracts including NSF Grants ATM-82-19020 and ENV 75-02667, and EPA Grant R-810403-01.

REFERENCES

1. Gordon, G. E. (1980) Receptor models. Environ. Sci. Technol. 14, 792-800.
2. Gordon, G. E., Pierson, W. R., Daisey, J. M., Lioy, P. J., Cooper, J. A. and Watson, J. G., Jr. (1984) Considerations of design of source apportionment studies. Atmos. Environ. (in press).
3. Henry, R. C., Lewis, C. W., Hopke, P. K. and Williamson, H. J. (1984) Review of receptor model fundamentals. Atmos. Environ. (in press).
4. Kowalczyk, G. S., Gordon, G. E. and Rheingrover, S. W. (1982) Identification of atmospheric particulate sources in Washington, D.C., using chemical element balances. Environ. Sci. Technol. 16, 79-90.
5. Andren, A. W., Klein, D. H. and Talmi, Y. (1975) Selenium in coal-fired steam plant emissions. Environ. Sci. Technol. 9, 856-858.
6. Pillay, K. K. S. and Thomas, C. C. (1971) Determination of the trace element levels in atmospheric pollutants by neutron activation analysis. J. Radioanal. Chem. 7, 107-118.
7. Tuncel, S. G., Olmez, I., Parrington, J. R., Gordon, G. E. and Stevens, R. K. (1984). Composition of fine particle regional sulfate component in Shenandoah Valley. Environ. Sci. Technol. (submitted).

8. Howes, J. E., Cooper, J. A. and Houck, J. E. (1984) Sampling and analysis to determine source signatures in the Philadelphia area. Draft report to EPA from Battelle Pacific Northwest Laboratory.

9. Ondov, J. M., Zoller, W. H. and Gordon, G. E. (1982) Trace element emissions on aerosols from motor vehicles. Environ. Sci. Technol. 16, 318-328.

10. Pierson, W. R. and Brachaczek, W. W. (1983) Particulate matter associated with vehicles on the road-II. Aerosol Sci. Technol. 2, 1-40.

11. Thurston, G. D. and Spengler, J. D. (1984) A quantitative assessment of source contributions to inhalable particulate matter pollution in metropolitan Boston. Atmos. Environ. (submitted).

12. Lewis, C. W. and Einfeld, W. (1984) Origins of carbonaceous aerosol in Denver and Albuquerque during winter. Environ. Intern. (submitted).

13. Rahn, K. A., Noelle, L. F. and Lowenthal, D. H. (1983) Elemental tracers of Canadian smelter aerosol transported into the northeastern United States. In Receptor models applied to contemporary pollution problems, S. Dattner and P. K. Hopke, ed. (Air Pollut. Contr. Assn., Pittsburgh).

14. Cooper, J. A., Currie, L. A. and Klouda, G. A. (1981) Assessment of contemporary carbon combustion source contributions to urban air particulate levels using carbon-14 measurements. Environ. Sci. Technol. 15, 1045-1050.

15. Currie, L. A. and Klouda, G. A. (1982) Nuclear and chemical dating techniques: Interpreting the environmental record. In Counters, accelerators and chemistry, L. A. Currie, ed., ACS Symp. Series No. 176, American Chemical Soc., Washington, D.C.

16. Parrington, J. R., Zoller, W. H., Olmez, I. and Gordon, G. E. (1984) Sources and composition of atmospheric particles of the free troposphere over the Pacific. J. Geophys. Res. (submitted).

17. Rahn, K. A. and Lowenthal, D. H. (1984) Elemental tracers of distant regional pollution aerosols. Science 223, 132-139.

18. Stevens, R. K., Dzubay, T. G., Lewis, C. W. and Shaw, R. W., Jr. (1984) Source apportionment methods applied to the determination of the origin of ambient aerosols that affect visibility in forested areas. Atmos. Environ. 18, 261-272.

19.   Heffter, J. L. (1983) Branching atmospheric trajectory (BAT) model, NOAA Tech. Memo ERL ARL-121.

20.   Shaw, R. W., Jr. and Paur, R. J. (1983) Composition of aerosol particles collected at rural sites in the Ohio River Valley. Atmos. Environ. 17, 2031-2044.

21.   Gluskoter, J. J., Ruch, R. R., Miller, W. G., Cahill, R. A., Dreher, G. B. and Kuhn, J. R. (1977) Trace elements in coal: occurrence and distribution. Ill. State Geological Survey Circular 499.

22.   Ondov, J. M., Biermann, A. H., Heft, R. E. and Koszykowski, R. F. (1981) Elemental composition of atmospheric fine particles emitted from coal burned in a modern electric power plant equipped with a flue-gas desulfurization system. In Atmospheric aerosol: source/air quality relationships, E. S. Macias and P. K. Hopke, eds., ACS Symp. Series No. 167, American Chemical Soc., Washington, D.C., Ch. 9.

23.   Ondov, J. M. (1984) private communcation.

24.   Daisey, J. M. (1980) Organic compounds in urban aerosols. Ann. N.Y. Acad. Sci. 338, 50-69.

25.   Friedlander, S. K. (1981) New developments in receptor modeling theory. In Atmospheric aerosol: source/air quality relationships, E. S. Macias and P. K. Hopke, eds., ACS Symp. Series No. 167, American Chemical Soc., Washington, D.C., Ch. 1.

26.   Daisey, J. M. (1984) A new approach to the identification of airborne mutagens. Environ. Intern. (submitted).

27.   Fehsenfeld, F. C. (1983) Field investigations of atmospheric chemical transformation and scavenging processes involving acid-related materials. Presented at the Review Meeting of the Nat'l. Acid Precipitation Assessment Program, Boston, Aug.

28.   Johnson, D. L., Davis, B. L., Dzubay, T. G., Hasan, H., Crutcher, E. R., Courtney, W. J., Jaklevic, J. M., Thompson, A. C. and Hopke, P. K. (1984) Chemical and physical analysis of Houston aerosol for interlaboratory comparison of source apportionment procedures. Atmos. Environ. (in press).

29. Johnson, D. L., McIntyre, B., Fortmann, R., Stevens, R. K. and Hanna, R. B. (1981) A chemical element comparison of individual particle analysis and bulk chemical analysis. Scanning Electr. Microsc. 1, 469-476.

30. Wieser, P., Wurster, R. and Seiler, H. (1980) Identification of airborne particles by laser induced mass spectroscopy. Atmos. Environ. 14, 485-494.

31. Davis, B. L., Johnson, L. R. and Flanagan, M. J. (1981) Provenance factor analysis of fugitive dust produced in Rapid City, South Dakota. J. Air Pollut. Contr. Assn. 31, 241-246.

32. Thompson, A. C., Jaklevic, J. M., O'Conner, B. H. and Morris, C. M. (1982) X-ray powder diffraction system for chemical speciation of particulate aerosol samples. Nucl. Instrum. Meth. Phys. Res. 198, 539-546.

33. Billings, C. E. and Matson, W. R. (1972) Mercury emissions from coal combustion. Science 176, 1232-1233.

34. Germani, M. S. (1977) Selected Studies of Four High-Temperature Air-Pollution Sources Ph. D. Thesis, Dept. of Chemistry, University of Maryland, College Park, MD.

35. Rahn, K. A. and Fogg, T. R. (1983) Boron as a tracer of aerosol from combustion of coal. Final Techn. Rep. on DOE Grant DE-FG22-82-PC51260, Grad. School of Oceanography, Univ. of Rhode Island, Narragansett, RI.

36. Kitto, M. E., Anderson, D. L., Gordon, G. E. and Ondov, J. M. (1983) Boron as a potential tracer for coal-fired power plant emissions. Presented at the American Chemical Society Nat'l. Mtg., Washington, D.C., Aug.

37. Braman, R. S. and Johnson, D. L. (1974) Selective absorption tubes and emission technique for determination of ambient forms of mercury in air. Environ. Sci. Technol. 8, 996-1003.

38. Ballantine, D. S., Jr. (1983) The implication of mercury speciation data from atmospheric studies on the global mercury cycle. Ph.D. Thesis, Dept. of Chemistry, Univ. of Maryland, College Park, MD.

39. Mosher, B. W. and Duce, R. A. (1983) Vapor phase and particulate selenium in the marine atmosphere. J. Geophys. Res. 88, 6761-6768.

40.   Gladney, E. S., Wangen, L. E., Curtis, D. B. and Jurney, E. J. (1978) Observations on boron release from coal-fired power plants. Environ. Sci. Technol. 12, 1084-1085.

41.   Lewis, C. W. and Stevens, R. K. (1984) Hybrid receptor model for secondary sulfate from a coal-fired power plant. Atmos. Environ. (submitted).

42.   Lewtas, J. (1984) The Integrated Air Cancer Project. Presented at the Workshop on Genotoxic Air Pollutants, Quail Roost, NC, April.

USE OF TRACERS FOR THE STUDY OF ATMOSPHERIC CHEMICAL AND PHYSICAL TRANSFORMATION PROCESSES

Stephen E. Schwartz

Environmental Chemistry Division
Brookhaven National Laboratory
Upton, NY 11973

ABSTRACT

Despite insights afforded by laboratory studies into chemical and physical transformation processes in the atmosphere, nonetheless it remains desirable to observe the actual occurrence of such transformations in the ambient atmosphere and/or to measure the rates of these processes. The basis of such studies is observation of changes in concentrations of species of interest as a function of time, but in essentially all cases this translates into a need to observe concentrations as a function of position, the conversion between these two variables being the transport wind velocity. However, concentration changes resulting from processes other than chemical or physical transformation can interfere with the interpretation of transformation processes that is the objective of the study. Potential sources of error include inaccuracy in following an air parcel (especially in the presence of a concentration gradient), concentration changes resulting from dilution, and/or failure of an underlying steady-state assumption. This paper examines the use of tracer compounds as a means of avoiding such errors. Applications of conservative tracers include use as a Lagrangian marker and/or as a concentration standard. Also considered are applications of reactive tracer compounds to the study of atmospheric transformations. Examples of these several types of application are presented.

I. INTRODUCTION

A major objective of research in atmospheric science over the past several decades has been to obtain an enhanced description of transport, transformation, and deposition of trace constituents present in the atmosphere. Interest in these processes has been heightened with the recognition that pollutant materials transported over long distances may exert significant effects at receptor locations and the consequent desire to improve knowledge of source-receptor relationships.

Much of the motivation for development and application of techniques employing tracer compounds in the atmosphere has arisen from the desire to understand transport and dispersion processes, and this sort of application has been the subject of several papers presented at this workshop. Tracers are employed in such application for the purpose of identifying particular air parcels and as a means of determining the amount of dilution that has taken place as a function of position and/or time. Tracer compounds useful for these purposes include both deliberately introduced materials and materials that are characteristic of a particular source or source region. Such studies may thus be distinguished from studies of transport and dispersion by the objective of characterizing the nature, extent, and/or rate of chemical or physical transformations. This objective may be in addition to the objective of determining transport and dispersion, or it may be the sole objective, transport and dispersion being mere complications to the study that must be sorted out, for which recourse is made to the use of tracers. In addition to such applications of "passive" tracers, we note as well the possibility of conducting studies of transformation processes by means of tracer compounds that undergo physical or chemical transformations that more or less closely resemble the processes pertinent to the species of interest. In studies with such "active" tracers, one makes use of a unique or known source distribution of the tracer, as well as its capability of measurement--in this case to ascertain the nature, rate, and/or extent of transformation, and perhaps as well of transport and dispersion.

This paper deals with the design of experiments employing tracer compounds for the study of atmospheric transformations and presents examples from the literature that embody one or another aspect of the design features discussed. Emphasis is given to the means of interpreting measurements to evaluate the extent and/or rate of the process or processes under examination.

II. BACKGROUND

Before turning to specific applications of tracers in study of transformation processes, it may be useful to address the question of why transformation processes should even be studied in the atmosphere. After all, it might be argued that the study of chemical transformation, specifically the study of reaction kinetics, is a well-developed and sophisticated laboratory science. Therefore, the argument runs, all that is necessary in order to describe transformation processes in the atmosphere is to study the pertinent processes in the laboratory to obtain a description of these processes that is sufficient to allow evaluation or prognostication of the rates of these processes in the atmosphere.

Laboratory investigation of atmospheric transformation processes can take place by one or both of two approaches. The first postulates that no matter how complicated a reaction system is, the mechanism of reaction consists of a sequence of elementary reactions, unimolecular, bimolecular, occasionally termolecular, that are characterized by very simple rate laws. Once these rate laws are determined, they may be extrapolated to permit rate evaluation for arbitrary reagent concentrations. This philosophy leads to an approach wherein reactions are individually studied under selected conditions that are best suited to determination of their rate laws. The description of a complex mixture of compounds and large number of reactions is then synthesized as the parallel and series combination of the elementary reactions. This approach has been widely applied and has led to the multi-species, multi-reaction mechanisms with which we are all familiar. Despite the philosophical appeal of such an approach, if this is the sole approach taken, questions must always remain as to whether all important processes have been included in the model.

As a contrast to the approach based on elementary reactions, and/or as a means of evaluating the validity of models based on such an approach, there is a second approach to the laboratory study of atmospheric transformations, which we denote the batch-reactor approach. By this approach, one introduces into a reactor a known mixture of reagents and turns on the chemistry (for example, by turning on the lights for photochemical reaction or by introducing an aerosol or dispersion of liquid water droplets for mixed-phase chemistry). The chemistry is followed by monitoring concentrations of selected species-- it is generally not possible to follow concentrations of all species. The dependence of the chemistry on parameters such as the mix of reagents, light intensity, and temperature can be determined and used to predict the chemistry of the ambient atmosphere. Because this approach is empirical, it is necessary to carry out studies over an appropriately wide range of conditions (most importantly, mixes of reagent concentrations) since extrapolations not based on mechanisms of elementary reactions are not necessarily warranted and can be erroneous.

An important link between the two types of approaches sketched here is comparison of the evolution of the batch-reactor composition with predictions of models built on elementary reactions. Such a comparison leads to confidence both in the completeness of the reaction mechanism and in the extrapolation of the results beyond the conditions measured in the reactor. Unfortunately, the comparison is generally not perfect. For example, in photochemical studies the transformations frequently "take off" faster than the model would predict. In such cases, the argument is made that there are possible "chamber effects," e.g., heterogeneous chemistry on reactor walls that produces some nitrous acid that photolyzes to give an "early" OH radical that gets the chemistry going. However, after suitable modification of the model to take such chamber effects into account, there is more or less good agreement with model predictions.[2,3] Nonetheless, there still remain questions about mass balance, e.g., missing NO_x--did it go to the wall?--etc.

It is out of concern about such possible deficiencies in the laboratory approach that the temptation rises to study chemical reactions in the atmosphere. Additionally, one always has the lurking suspicion that something important may have been overlooked in the chamber study (e.g., highly reactive natural hydrocarbons) that may substantially alter the chemistry, although to a very great extent that sort of objection can be overcome by filling the chamber with ambient air.[4] Nonetheless, it may still be argued that a laboratory study cannot adequately simulate the atmospheric process of dispersion and mixing. For these reasons and perhaps also because a field measurement of transformation rates is perceived as more convincing than the laboratory approach, we are faced with the desire to carry out studies of chemical transformations in the ambient atmosphere.

The objective of this paper is to examine the use of tracers, including tracers of opportunity, in the conduct of such studies, to examine methods of interpreting such studies, and to point out difficulties and ambiguities in such interpretation.

III. APPROACHES TO EXPERIMENTAL DESIGN

Having accepted the need or desire to conduct studies of transformations in the ambient atmosphere, we are now faced with the question of how to conduct such studies. Perhaps the simplest approach might be to set up a station to measure the concentrations of a few species of interest and follow the time evolution of these concentrations at such a station. Then, depending on one's boldness or naiveté, one ascribes measured change in concentrations to the occurrence of transformation processes. The problem with this approach is that it neglects transport--a neglect that prompted one researcher to express the "hope... that eventually it will be recognized that, sometimes, air moves."[5] In brief, changes in species concentrations may be due not only to chemical reactions but also (and perhaps to a greater magnitude) to advection in the presence of concentration gradients, or to dilution (for example, associated with an increase in the height of the mixed layer).

An alternative approach is to take explicit advantage of the fact that air moves. By analogy to laboratory kinetic studies with flow-tube reactors, one devises a study in the atmosphere wherein the concentration of a species is measured at one location and then again at a second location downwind. The transit time between the two points is evaluated from the wind speed, and from changes in composition, chemical kinetic rates may be evaluated. There are several possible pitfalls associated with this approach that are well known to most of us present at this workshop, but which may nonetheless be useful to review here. First, this approach can assume that the measurements are carried out in a Lagrangian mode;[6] that is, the second, downstream measurement is made on the same air parcel as the upstream measurement, the identification of the parcel having been made by some sort of marker or taggant or by wind-speed and trajectory analysis. Alternatively, it can be assumed that the system is at steady state, so that the parcel sampled at the downstream location is representative of the parcel sampled at the upstream location. Failure to satisfy either of these conditions can again result in falsely ascribing to transformation changes in concentrations caused by advection. A further and very real concern is dilution. Consider measurements made "near" a source so that the concentration profile is not uniform in the y and z directions (the x direction is the direction of the mean wind); the most obvious example would be a point-source plume. It is clear that as one goes downwind, the concentrations of various plume constituents will decrease simply as a consequence of turbulent dispersion processes, and consequently, that proper account of concentration changes caused by dilution must be taken in the design and interpretation of transformation studies in the ambient atmosphere, based on following an air parcel or on distance-to-time conversion. Also, since the air which is diluting the plume may not contain a negligible concentration of the species of interest, proper account must be taken of the upwind or background component of the measured concentrations that results from plume dilution. Indeed, the process of measuring kinetic rates in the atmosphere from variation in concentration in the direction of the mean wind has been likened to the use of a flow tube apparatus in the laboratory, except that the reagent concentrations are poorly characterized, the system may not be at steady state, the flow speed is not well established and the walls are porous.

It is to help overcome the difficulties of studying chemical and physical transformations in the atmosphere that one turns to the application of tracer compounds. To facilitate discussion, it is useful to distinguish five different types of applications of tracer compounds in the study of atmospheric transformations.

- Type 1, Tracer as Plume Marker. The use of a tracer to identify a plume of air so that it may be distinguished from surrounding unmarked air with the objective of following changes in composition of this plume as a function of downwind distance.

- Type 2, Tracer as Lagrangian Marker. Similar to Type 1, but with the further specific objective of allowing an air parcel to be followed in a Lagrangian sense, thereby diminishing uncertainty arising from steady-state treatment or from identification of the Lagrangian parcel solely by means of wind speed. Specifically included in this category are studies employing balloons as tracers.

- Type 3, Tracer as Concentration Standard. The use of a tracer compound (ideally a conservative or nonreactive compound) as a concentration standard against which to normalize the effects of dilution. Insofar as it may be argued that the tracer and the compound or compounds of interest experience the same advection and dilution, the effect of dilution may be eliminated by ratioing concentrations to the tracer concentration. In this category one should perhaps include studies of chemical reactions in plumes, for which some plume constituent (or sum of plume constituents) can be considered a conservative tracer; the extent of reaction is followed by ratio to the concentration of this conservative tracer.

- Type 4, Tracer as Clock. The use of a tracer undergoing a transformation process at a known rate as an internal clock against which to measure the rate of some other process, e.g., dispersion, transport, or transformation.

- Type 5, Tracer as Reactant. The use of a chemically reactive or physically scavengeable compound whose chemistry more or less closely mimics that of the compound of interest. From an examination of the fate of the tracer and/or the kinetics of the process, information may be gained pertinent to the compound of interest. As an example, one might consider an experiment with a tracer that reacts with an OH radical with known rate constant. From the disappearance rate, one infers the OH radical concentration and, in turn, the rate of reaction with OH of other compounds of interest. A second example would be the use of isotopically labeled compounds as reactive tracers. The isotopically labeled compound, of course, mimics closely all the transformation properties of the compound of interest yet retains an isotopic signature that eliminates interference from the existing concentrations of the nonlabeled compound, permitting study of transformation processes in the exact atmospheric milieu experienced by the compound of interest.

Although the foregoing types of applications can be considered logically distinct, it should be noted that a given experimental design may embody more than one of these types of application. Nonetheless, we find it useful to retain the distinction between the several types of application in examining various hypothetical or actual experimental designs.

IV. DESIGN, INTERPRETATION, AND EXAMPLES

In this section, we consider various models of conducting and interpreting tracer experiments to yield information about atmospheric transformation processes. Representative examples of the application of the several types of experimental design are presented. These examples are selected as being illustrative of the several approaches, with no attempt made to present an exhaustive literature survey.

A. Type 1, Tracer as Plume Marker

Experiments within this category rely on the identification by a tracer of a plume in which the evolution of composition is to be examined; the plume results from the release of the tracer over an extended period of time. The tracer serves to distinguish the marked sample from the surrounding air in which it is embedded to facilitate comparison of composition in successive measurements. Marker studies may also comprise a component of a more comprehensive investigation, thus fitting also into one or more of the categories outlined below. Kinetic information may be derived from the distance-to-time conversion, making use of the measured wind speed. An attempt may be made to follow a given parcel within the plume, making use of wind speeds in real time (or postanalysis of sampling conducted at numerous locations) to infer composition changes within a particular air parcel. Alternatively, no explicit attempt may be made to carry out the measurements and/or analysis in a Lagrangian frame and the system (source as well as atmospheric transformations) is treated as if at steady state. In addition to studies conducted with deliberately introduced tracers, it may be useful also to consider within this category studies of the evolution of chemical composition as a function of distance (i.e., age) within a plume either of a point source or of a distributed source (i.e., urban plume), the plume itself (or some constituent of the plume) constituting a self-tracer. As examples of such studies, reference is made to work of Pueschel and colleagues (e.g., Refs. 7 and 8), who have carried out numerous investigations in a variety of point source plumes, examining such processes as cloud nucleus formation, SO_2 to $SO_4^=$ conversion, etc.

An example of the use of a deliberately added tracer to distinguish a particular source is shown in Figs. 1-3. In this study,[9,10] a single source (the Eggborough power station) was labeled by continuous SF_6 injection, and it was possible thereby to distinguish the plume of this particular source from other sources to a distance of 650 km, corresponding to a travel time of approximately 24 h. Notable in this particular instance were the substantial fraction of NO remaining unoxidized and the corresponding persistent ozone depletion, both of which would appear to be related to the low degree of dispersion indicated by the tracer profile.

Also falling within this category are studies making use of tracers of opportunity. For example, Macias et al.[11] have noted increases of aerosol sulfate associated with aerosol lead transported from southern California. These authors were able also to pinpoint the location of the SO_2 to $SO_4^=$ conversion because the size of the sulfate-containing particle depends on the relative humidity at the time of oxidation, larger particles being characteristic of higher relative humidity. Several investigations have noted the higher ozone-forming potential of urban air, which is characterized by a high concentration of fluorocarbon-11 (F-11) relative to nonurban background air.[12-14] Likewise, the so-called Arctic haze has been linked to secondary aerosol formation by association either with particular trace metals[15] or trace gases[16] transported from source regions. It has been argued as well[17] that ratios of trace metals (Mn/V) can be used to pinpoint the origin of secondary sulfate aerosol, in particular that the Northeast U.S. is the source of much of its own sulfate. However, in rebuttal it has been noted that in situations where a species employed as a tracer of opportunity may derive from more than a single source region, the presence of the tracer (or a particular ratio of one tracer to another) does not uniquely pinpoint the source of the associated pollutant.[18] In general, extreme caution must be exercised in any claim of "guilt by association" of a pollutant with a tracer of opportunity.

B. Type 2, Tracer as Lagrangian Marker

Here the ideal experiment consists of an instantaneous release of tracer to mark an air parcel the evolution of whose composition is to be followed. The essential difference between plume marker studies (Type 1) and Lagrangian marker studies (Type 2) is the objective of using the tracer to identify a particular air parcel whose composition is to be examined as that parcel is advected. The tracer may consist of a gas or aerosol or alternatively, as noted above, of one or more balloons. Use of a gas or aerosol allows determination of the extent of atmospheric dispersion of the parcel containing the species of interest and of the volume over which sampling must be conducted in order to obtain a representative sample. In practice, the extent of this dispersion may not be well characterized. Instantaneous release of the marker provides unambiguous time resolution, but locating the puff may present difficulties. One resolution of this difficulty has been the occasional release of a puff of a second tracer in conjunction with continuous release of a primary tracer (Fig. 4). Another possible approach might be the sustained simultaneous release of two tracer compounds, one at a constant rate and a second at a rate that is ramped as a function of time, as indicated in Fig. 5. From the ratio of the concentrations of the two tracers, an unambiguous measure might thus be obtained of the elapsed time between release and sampling. Lagrangian markers may also, under fortuitous circumstances, be found in tracers of opportunity. Some years ago, Rasmussen noted an impulse of F-11 concentrations in an urban plume resulting from early morning household use of products containing these compounds as propellants and suggested that this impulse could be used as a time marker of opportunity imbedded within urban plumes. However, in practice this approach has not proved to be of very great utility.[19]

Constant level (or constant density) balloons have been employed as Lagrangian markers in quite a few studies of chemical transformations; for reviews see Clarke et al.[20] and Zak.[21] Reference 20 presents an overview of the use of tetrahedral balloons (tetroons) as Lagrangian markers in long-distance (250 to 1,000-km) transport studies; see Fig. 6. The tetroon may be followed visually or preferably by means of a radar transponder; the latter requires cooperation of aviation authorities. Air sampling for the purpose of following chemical transformation may be achieved by aircraft or by ground vehicles. Reference 21 also describes the application of a manned balloon for atmospheric chemistry studies (Project Da Vinci). The use of super-pressure balloons carrying a compact, lightweight payload of automated sampling equipment has also been proposed.[22]

Limitations of balloons as Lagrangian markers are discussed in some detail in Ref. 20. These limitations arise in the first instance from the design property that the balloon follow a surface of constant density as opposed to air parcels, which, except in the vicinity of large-scale precipitation, are generally transported on isentropic surfaces (constant potential temperature). Additionally, balloons are subject to changes in their effective buoyancy as a result of solar heating, radiative cooling, stretching of materials, leakage, and dew condensation that limit their ability to follow a constant pressure surface. The altitude variation in a particular tetroon trajectory is shown in Fig. 7. Notable here are effects of convective updrafts during the day and most likely dew formation at night.

Even considering the many problems associated with use of tetroons, Clarke et al.[20] conclude that they are adequate Lagrangian plume markers in the daytime boundary layer. However, they caution that at night, when large wind shears typically exist, a small variation in the altitude of the tetroon may lead to a significant change in trajectory.

An extensive treatment of a case study of ozone formation in conjunction with long-range transport as studied with assistance of a tetroon marker is given by Clark and Clarke.[23] In this study, cross-plume transects were conducted by aircraft at a variety of locations as the tetroon was transported from Baltimore, Maryland, to southwestern Connecticut (Fig. 8); species measured included O_3 and NO_x (real-time continuous) and hydrocarbons (grab samples). In this study, it was possible to follow O_3 formation in the Baltimore-Washington plume and, as well, to discern the effects of incremental pollutants added as the parcel incorporated emissions from Philadelphia (Fig. 9) and New York. It was found that the O_3 concentration within the urban plume increased approximately 100 ppb during daylight hours, whereas the increase outside the urban plume amounted to only about 30 ppb.

Another valuable study utilizing tetroons as Lagrangian markers was the Los Angeles Reactive Pollutant Program (LARPP), as described by Calvert.[24] Fig. 10 shows the time dependence of O_3, NO, and NO_2 during the course of a 6-hour study; sampling was conducted by helicopters following the tetroon. Evident in the ozone data are an initial value of 50 ppb, followed by a decrease as newly emitted NO was mixed into the parcel. Later in the day, the O_3 concentration increased concurrently with a decrease in the]NO[/]NO$_2$[ratio (Fig. 11).

A study of sulfate formation in a source-free region was reported by Alkezweeny,[25] who monitored sulfate concentrations using a tetroon as marker over Lake Michigan (Figs. 12 and 13).

A major obstacle to deriving quantitative information about chemical and physical transformation rates from concentration measurements, even if the air parcel is being followed in a Lagrangian manner, is the difficulty in distinguishing the change in concentration caused by transformation (either decrease in reagent concentration or increase in product concentration) from that due to dispersion and perhaps other loss processes, including dry deposition. As discussed below, substantial advance may be made by ratioing the concentration of the species of concern to that of a tracer. We note here, however, that in general it is not appropriate to ratio to the concentration of a tracer employed as a Lagrangian marker, because this species (ideally released as a puff) would be dispersing much faster than species of concern (unless both were released together).

C. Type 3, Tracer as Concentration Standard

We have noted that in order to evaluate the extent or rate of a transformation process of concern from the measured total change in species concentration, it is necessary to account for changes in the species concentration by other processes. A large and generally dominant change in concentration is that caused by dilution. This dilution may result from lateral and vertical dispersion near the source or, for a situation sufficiently far from a source that the substance of concern has flooded the mixed layer, may result from an increase in the height of the mixed layer. A very valuable application of tracers is to evaluate the concentration change due to dilution and thereby permit determination of the change in concentration resulting from the transformation process of interest. More specifically, the effect of dilution can be accounted for by ratioing the concentration of the species of interest to that of the tracer, provided that both species experience the same dilution.

The potential source of error in this approach results from the species of interest and the tracer not being emitted from the same point, as illustrated schematically in Fig. 14. Here the two locations 1 and 2 present effective source points for the two species. It is evident that for the two source points not co-located, the ratio C_2/C_1 decreases with age of distance simply as a consequence of the differing rates of dilution. In my judgment, this effect places serious limitation on the use of a tracer from a non co-located source as a dilution standard. This observation would thus apply prima facie to any proposed use of a non co-emittted tracer of opportunity, with the burden of proof of the suitability being on the proponent. On the other hand, measurement of change in the ratio of concentrations of substances co-emitted at constant, known ratio would appear to be well suited to determination of the extent or rate of atmospheric transformation processes.

Measurement of change in concentration ratios of co-emitted substances is, of course, the basis for interpretation of transformations occurring in plumes. Here the source may be a single point source or alternatively may be an urban area. The tracer employed in such analyses is most commonly a co-emitted tracer of opportunity, although a few studies have been conducted with intentionally introduced tracers. Most commonly the system has been treated as a steady-state reactor, although some attempt may be made to conduct measurements in a near-Lagrangian mode, making use of measured wind speeds. Occasionally studies have been conducted with Lagrangian markers.

Considerable work has been carried out in the last 15 years on SO_2 oxidation in stack plumes, and thus it seems worthwhile to refer to such studies to illustrate the features of this type of study. For reviews of this research see Newman,[26] Wilson,[27] and Levy et al.[28] Attention is called also to papers presented at the 1980 Symposium on Plumes and Visibility.[29] For the most part in these studies, SO_2 is considered a self-tracer, and the extent of reaction is evaluated from the change in the ratio $]SO_4^=[/(]SO_2[+]SO_4^=[)$, total sulfur (i.e., $]SO_2[+]SO_4^=[)$ being assumed to be a conservative tracer. In principle the ratio should represent a cross-sectional average over sulfur in the plume,[30,31] but for a well-mixed plume (or for oxidation assumed independent of y and z), point measurements or ratios over transverses appear adequate. In place of $]SO_4^=[$ a surrogate such as b_{scat} may be employed, with suitable conversion from b_{scat} to $]SO_4^=[$. Alternatively, rates may be expressed directly, for example, as an Aitken nucleus production rate normalized to SO_2 concentration in the plume.[32] In principle, the extent of reaction might be followed as a decrease in the ratio $]SO_2[/(]SO_2[+]SO_4^=[)$, but for small extent of reaction this is less sensitive.

The assumption that $]SO_2[+]SO_4^=[$ is conservative would appear to be valid in stack plume studies prior to contact of the plume to the surface, after which time SO_2 removal (and perhaps also $SO_4^=$ removal) by dry deposition must be accounted for.[30,33,34] The rate of SO_2 oxidation may be evaluated from the increase in $]SO_4^=[/(]SO_2[+]SO_4^=[)$ with age and is commonly reported in units of per cent per hour. In most studies, an oxidation rate of up to several per cent per hour has been found, with indication of variation in this rate with insolation. In some instances, it has been possible to infer a reaction order (in plume constituents) higher than first order, suggesting possible interaction of SO_2 with co-emitted particulate matter;[35,36] the kinetic analysis takes explicit advantage of dilution within the plume, which can be measured by the tracer.[37] Analysis of kinetics in expanding stack plumes indicates that higher than first-order reactions may be quenched by plume dispersion.[37,38]

A few studies of SO_2 oxidation in plumes have employed tracers other than sulfur itself. Eatough and colleagues[39,40] studying reactions in a smelter plume employed co-emitted arsenic as a conservative tracer. These authors were able to study not only SO_2 oxidation but also formation of metal-stabilized S(IV) complexes and neutralization of plume acidity as the plume incorporated ambient basic material. Analysis of the kinetic rate law in terms of plume expansion rates established that the oxidation of SO_2 was a first-order process. In other studies,[41,42] Eatough and his colleagues have used as tracers co-emitted Ni, V, and total NO_x ($NO + NO_2 + HNO_3 + PAN$) as tracers. In Ref. 42 results were obtained for the plume encountering a fog bank, for which SO_2 oxidation rates of 30% h^{-1} were reported.

An intriguing study that to my mind has not been adequately resolved is the early study of Weber,[43] who reported measurements of SO_2 and CO_2 excursions above baseline when the plume of a local power station (distance several km) impacted his measuring equipment. A portion of Weber's data is reproduced in Fig. 15. From the ratio of SO_2 to CO_2 concentrations above baseline in comparison to emissions rates, Weber calculated SO_2 removal rates of 50 to 300% h^{-1}, depending on meteorological conditions. The reason for the rather large removal rate (in comparison to rates measured in later plume studies) is not apparent. One possibility is that surface sampling of SO_2 is highly influenced by ground-level SO_2 removal and is thus not representative of losses from the entire plume. In this conjunction it might be observed that substantial differences between elevated (100 and 200 m) and ground level SO_2 concentrations have been routinely observed in measurements in the Netherlands.[44]

A few measurements of SO_2 loss from stack plumes have been reported from studies where a conservative tracer (SF_6) has been deliberately introduced into the stack as a concentration standard. These measurements have not in general been very successful, although, based upon experience[45] using SF_6 as a plume tracer (Fig. 16), the likelihood of successfully using SF_6 as a concentration standard would appear to be high. In an early study of a coal-fired power plant plume, Dennis et al.[46] found an apparent rapid initial decrease in the $]SO_2[/]SF_6[$ ratio (~25%) compared to the ratio in the stack. The $]SO_2[/]SF_6[$ ratio then decreased more slowly at a rate of ~50% h^{-1}. Flyger and colleagues,[47,48] studying an oil-fired power plant, found a half-life of ~30 min (SO_2 loss rate ~140% h^{-1}). Easter et al.[31] report measurements of both $]SO_2[/]SF_6[$ and $]SO_4=[/]SF_6[$ ratios in one oil-fired and two coal-fired power plants. Reported SO_2 loss rates relative to SF_6 were occasionally rather high (several tens of per cent per hour) but were not uncommonly negative, indicating substantial measurement error associated with this technique. $]SO_4^=[/]SF_6[$ data also showed considerable scatter, whereas analysis based on $]SO_4^=[/(]SO_4^=[+]SO_2[)$ ratio gave a much more consistent picture of SO_2 oxidation rates. A possible explanation for the variability in ratios to $]SF_6[$ may be the existence of large short-distance inhomogeneity in concentrations within the plume and the consequent sampling by the different instruments of differing amounts of "plume" and "nonplume" air. This problem is obviated by instrumentation such as filter packs, in which both the species to be measured ($SO_4^=$) and the tracer (SO_2) are determined in exactly the same volume of air.

Several studies have been conducted of SO_2 to $SO_4^=$ conversion in urban plumes by using total sulfur as a self-tracer.[25,49-51] In these studies, dilution due to increasing height of the mixed layer may be explicitly accounted for, but it is necessary to estimate the effects of SO_2 loss due to dry deposition. SO_2 oxidation rates of 0 to 12% h^{-1} are reported. Recently, Lewis and Stevens[52] have developed an extension of the receptor modeling approach that incorporates SO_2 to $SO_4^=$ conversion, making use of an assumed emission ratio of SO_2 to fine particles characteristic of coal-fired sources.

In comparison with tracer-based studies of SO_2 oxidation, relatively fewer studies have been conducted of nitrogen oxide chemistry, despite (or perhaps because of) the richer chemistry of the nitrogen oxides. This chemistry includes oxidation of NO to NO_2 by ozone and free radicals, net ozone production, and further oxidation of NO_2 to HNO_3 and peroxyacetyl nitrate (PAN). A very insightful analysis of this chemistry was presented by White,[53] who treats the chemistry in terms of what he denotes "invariants," conserved quantites. For example, for NO emitted into a background of O_3, the rapid reaction

$$NO + O_3 \ \ell \ NO_2 + O_2$$

produces a local ozone deficit that is diminished as the plume disperses into the background. White points out that treatment of this process and examination for net ozone formation is simplified by considering the sum $]NO_2[+]O_3[$, which would be constant (and equal to the background O_3 concentration + primary $]NO_2[)$ in the absence of more interesting reactions such as net O_3 formation or NO_2 oxidation. Similarly, the sum $]NO[+]NO_2[$ is a quasi-conservative tracer, like SO_2. In his treatment of this chemistry, White, in fact, models the dispersion by treating SO_2 as a plume tracer. Based on this treatment, White was able to show the absence of net O_3 formation for a particular set of measurements in a coal-fired stack plume. In contrast, other studies, e.g., Ref. 54, have clearly demonstrated net O_3 formation in stack plumes. One further study to which attention should be called in the present context is that of Easter et al.,[31] in which SF_6 and SO_2 were used as conservative tracers against which to examine for NO_x loss. The scatter in the measurements was quite large, however, and little was obtained in the way of definitive results.

A number of studies have been conducted of NO_x reactions in urban plumes utilizing a variety of tracers of opportunity. Calvert[55] presents data for loss of NO_x (= NO + NO_2 + PAN + HNO_3, where is an efficiency for detection of HNO_3, 0 1) in a study that was conducted with a tetroon as Lagrangian marker in Los Angeles. C_2H_2 and CH_4, assumed to derive from similar source distribution as NO_x and in constant emissions ratio, were employed as concentration standards. The rate of loss of NO_x determined in this way was ~10% h^{-1}. Spicer[56] has examined the NO_x removal rate in the urban plume of Boston over Massachusetts Bay by using CO, C_2H_2, and F-11 as measures of dilution; the NO_x removal rate ranged

from 14 to 24% h^{-1}. Chang et al.[57] have examined both the ratio
$(]NO[+]NO_2[/]CO[)$ and $(]PAN[+]HNO_3[)/(]NO[+]NO_2[+]PAN[+]HNO_3[)$
to examine NO_x reaction rates. A number of other studies based on ratio to
CO or C_2H_2 have been discussed by Spicer[58] and Anderson.[59] Both authors
point out the sensitivity of the evaluated reaction rates to the assumed
emission inventories of the species.

By a treatment similar to that employed for nitrogen oxides Calvert[55]
has examined the rate of reaction of several reactive hydrocarbons (C_2H_4,
C_3H_6, 1-C_4H_8, iso-C_4H_8, n-C_5H_{12}, iso-C_5H_{12}); see Fig. 17. From the
inferred rate of reaction of these compounds (as well as NO_x) and known OH
rate constants, estimates were obtained for the OH radical concentration.
The range of these estimates for the morning hour of a particular day over a
path defined by tetroon markers with Lagrangian sampling was (0.5 -
4.3) x 10^6 cm^{-3}. The wide range in evaluated OH concentrations is evidently
a measure of the cumulated errors associated with determination of the loss
rates of the several species.

A similar treatment was made by Roberts et al.[60] of concentration
ratios of several aromatic compounds measured at a Colorado mountain site
under conditions of westerly flow. Making use of known ratios of these
compounds in urban emissions and back trajectories to evaluate transit times
from putative urban sources, the authors were able to estimate diurnal-
average OH concentrations; for the three compounds studied, these values
were (0.5-1.7) x 10^6, (0.2-1.8) x 10^6, and (0.35 - 1.25) x 10^6 cm^{-3}. Based on
consideration of uncertainties, an upper-limit diurnal-average HO concentration for rural air was given as 2.4 x 10^6 cm^{-3}.

Fehsenfeld et al.,[61] considering benzene to be a conservative tracer of
urban (Denver) air transported to a Colorado mountain site have been able to
estimate the amount of urban O_3 transported to that site. On this basis they
rule out transported urban O_3 as the cause for the high O_3 level observed at
that site and conclude that O_3 production is occurring either at that site or in
transit from the urban area.

D. Type 4, Tracer as Clock

Here measurement is made of the change in the ratio of concentrations
of two tracers that undergo transformation at known rates.

From the change in concentration ratio between two points, the elapsed time between the two points may be inferred. In turn, a change in concentration ratio of a species of interest to one or the other of the clock tracers can be used to evaluate the rate of transformation of that species. As may be inferred, the method is sensitive to the requirement that the three species have the same source distribution and, as well, requires that the lifetimes of the clock species be comparable to that of the species under investigation.

I am aware of only one application of this method. Nguyen et al.,[62] measured concentrations of ^{222}Rn and ^{212}Pb as well as SO_2 and $SO_4^=$ along the French Mediterranean coast, inland, and at sea. ^{222}Rn and ^{212}Pb are terrestrially derived radioisotopes having half-lives of 3.9 and 0.44 days respectively. Plots of the concentration data, according to the treatment given by Nguyen et al., are shown in Fig. 18. This treatment predicts a straight-line plot if the assumptions of the model are fulfilled. The treatment yielded a loss rate for SO_2 of 5.8% h^{-1}. The authors note that the mean transport velocity, which may also be evaluated from their treatment, agreed with the measured wind speeds.

In view of the stringent requirements on application of this technique, the technique is probably only of limited utility. Nonetheless, in my judgment, it should be given serious consideration in the arsenal of tracer techniques.

E. Type 5, Tracer as Reactant

This is the most direct use of tracers to study chemical and physical transformations. Release a material that closely simulates the substance of interest in its tendency to undergo chemical or physical transformations but which can be distinguished from the substance of interest, and follow the transformations of this material in the ambient atmosphere. These transformations can in principle be followed as a decrease in reagent concentration (relative to a concentration standard) or increase in product concentration (again relative to concentration standard) or may merely consist of identifying the pathways followed by the tracer (e.g., uptake by clouds or precipitation). In a sense, perhaps, various studies of reactions in plumes might be considered to fall into this category, where the substance of interest, e.g., SO_2, could be considered a reactive tracer of opportunity, since much of the motivation for the study of SO_2 reactions in stack plumes has derived from the desire to study the reactions of SO_2 in the atmosphere, and stack plumes are a convenient medium for such study. However, I choose not to include such studies within the category of reactive tracer studies and restrict consideration here to deliberately introduced reactive tracers. Nonetheless, essentially the entire discussion above that is pertinent to the user of conservative (or quasi-conservative) tracers to the study of transformations of substances normally present in the atmosphere (i.e., materials other than deliberately introduced reactive tracers) is pertinent as well to studies conducted with reactive tracers. In particular, the use of marker and concentration-standard methodologies will be of great utility in such studies.

A variety of aerosol tracers of rare earth metals and fluorescent organic dyes have been employed in studies of entrainment by clouds and precipitation. Important properties of these tracers are their size and solubility. The present status of technology in generating such aerosols is briefly reviewed in Ref. 63. Attention is called also to several papers in the Proceedings of the Precipitation Scavenging Conference of 1974, including those of Gatz,[64] Davis and Young,[65] Young et al.,[66] and Dingle.[67] Consideration may also be given to use of soluble gases as tracers. The important property of such tracers is their solubility, which controls the equilibrium of the gas between the two phases. The physical solubility of a gas in a liquid is given by Henry's law, which can be written for a species X as

$$]X[= H_X p_X;$$

for p_X in atm and $]X[$ in mole/liter (M), the Henry's law coefficient has units $M\ atm^{-1}$. The ratio of material present in aqueous phase to gas phase in any volume of space is

$$\frac{C(aq)}{C(g)} = \frac{L]X[}{p_X/RT} = LH_X RT ,$$

where L is the volume fraction occupied by liquid water, R is the universal gas constant, and T is the absolute temperature. For a gas to be distributed largely in cloudwater, it is seen that the Henry's law coefficient must substantially exceed $(LRT)^{-1}$, which for a cloud of liquid water content $L = 1 \times 10^{-6}$ (1 cm^3/m^3) has value $\sim 4 \times 10^4$ M atm^{-1}. This value is quite high and is approached or exeeded only by a few gases[68] (e.g., H_2O_2, $H = 1 \times 10^5$ M atm^{-1}). In view of the limited physical solubility of gases, attention is turned to gases that react in water, thereby increasing their solubility and distribution into water. Senum[69] has pointed out that quite a few aldehydes and ketones undergo hydration, increasing their solubility many-fold. For example, he notes that perfluoroacetone exhibits an equilibrium constant for hydration of about 1×10^6; the Henry's law solubility of this compound is apparently not known so that the overall solubility cannot be evaluated. It is, of course, this overall solubility that would govern the suitability of such a species as a tracer for scavenging of gas-phase species. Senum also notes that the suitability of a given species as a tracer for cloud scavenging will depend on the kinetics of hydration as well as on considerations of detectability, freedom from interference, and cost.

It is possible also to consider experiments with tracers having reactivities selected to mimic the tendency for particular gas-phase reactions of the species of interest. For example, since OH reactions dominate the gas-phase chemistry of a number of components of interest, one might consider experiments with a tracer gas selected to have similar OH rate constant to the species of interest. Calvert,[55] Roberts et al.,[60] and Altshuller[70] have discussed the relative reactivities of a variety of organic compounds with OH, O_3, and atomic oxygen. Again, suitability of a given compound as a reactive tracer would depend also on considerations of detectability, background, and cost.

A particular class of compounds that would potentially represent ideal reactive tracers are isotopically labeled analogs of the species of interest, since these tracers would accurately mimic all of the transformation properties of the species. However, although such isotopic tracers are highly attractive in that respect, their use as tracer compounds presents a new set of problems.

Two classes of isotopic tracers may be identified. The first would make use of an altered ratio of naturally occurring (stable) isotopes; collection of a sample and determination of the isotope ratio in that sample (by mass spectrometry or, perhaps, by an optical technique) would allow determination of the contribution of the isotopically altered tracer to the total sample. The principal limitation of this approach is that the variabilty of isotopes in normal emissions of the material of interest requires a high release rate of the isotopically altered compound, which is inevitably quite expensive. A second limitation is the possiblity of isotope fractionation of the tracer, which would tend to blur the isotopic signature. A third limitation is the difficulty and expense of sample preparation and handling for isotopic analysis.

A second approach, which seems to be technically feasible, is the use, where available, of a radioactive tracer, e.g., $^{35}SO_2$ for SO_2. The low natural abundance and high detectability of radioisotopes both stand in favor of their use as tracers. Additionally, isotope fractionation becomes a small error term rather than a loss of signal, as is the case with stable isotope tracers. However, restrictions on the deliberate release of radioisotopes may preclude their use as tracers. Such a restriction in fact precluded a study in Denmark that had intended to make use of $^{35}SO_2$ as a tracer.[47,71]

V. SUMMARY AND CONCLUSION

This paper has reviewed methodologies pertinent to the use of tracers for the study of chemical and physical transformations in the atmosphere. Several distinct approaches have been identified and their strengths and limitations outlined. In a few well-chosen situations, e.g., O_3 formation, Lagrangian studies carried out in conjunction with markers have yielded fairly convincing demonstrations of the extent of atmospheric reaction. However, in many instances, e.g., stack plume measurements of SO_2 oxidation by ratio to SF_6, or evaluation of OH concentration from Lagrangian determination of reaction rates of various hydrocarbons, the measurement errors are quite large. In such cases it may still be necessary to rely upon transformation rates derived on the basis of laboratory studies.

REFERENCES

1. Bufalini J. J., Walter T. A., and Bufalini M. M. (1977). Contamination effects on ozone formation in smog chambers Environ. Sci. Technol. II, 1181-1185.

2. Duewer W. H., Walton J. J., Grant K. E., and Walker H. (1978) Livermore Regional Air Quality (LIRAQ) Application to St. Louis, Lawrence Livermore Laboratory, Report UCRL-52432.

3. Whitten G. Z. and Hogo H. (1977). Mathematical modeling of simulated photochemical smog, Environmental Protection Agency, Report EPA-600/3-77-011.

4. Kamens R. M., Jeffries H. E., Sexton K. G., and Weiner R. W. (1982). The impact of day-old dilute smog on fresh smog systems: An outdoor chamber study. Atmos. Environ. 15, 1027-1034.

5. Slinn W. G. N. (1982). Letter to Participants, 4th International Conference on Precipitation Scavenging, Dry Deposition, and Resuspension.

6. Zak, B. D. (1983). Lagrangian studies of atmospheric pollutant transformation, in Advan. Environ. Sci. Technol. 12, S. E. Schwartz, Ed., Wiley, New York, pp. 303-344.

7. Pueschel R. F. and Valin C. C. V. (1978). Cloud nucleus formation in a power plant plume. Atmos. Environ. 12, 307-312.

8. Mamane Y. and Pueschel R. F. (1980). Formation of sulfate particles in the plume of the four corners power plant. J. Appl. Meteorol. 19, 779-790.

9. Cocks A. T., Kallend A. S., and Marsh A. R. W. (1983). Dispersion limitations of oxidation in power plant plumes during long-range transport. Nature, 305, 122-123.

10. Kallend A. S. (1983). The Fate of Atmospheric Emissions Along Plume Trajectories Over the North Sea - Summary Report. Electric Power Research Institute Report EA-3217.

11. Macias E. S., Zwicker J. O., and White W. H. (1981). Regional haze case studies in the southwestern U. S. - II. Source contributions. Atmos. Environ. 15, 1987-1997.

12. Cox R. A. and Eggleton A. E. J. (1975). Long-range transport of photochemical ozone in north-western Europe. Nature 255, 118-121.

13. Spicer C. W., Joseph D. W., Sticksel P. R., and Ward G. F. (1979). Ozone sources and transport in the northeastern United States. Environ. Sci. Technol. 13, 975-985.

14. Jaffar, M., Dutkiewicz, V. A., and Husain, L. (1981). Trichlorofluoromethane as a tracer of urban air masses. Atmos. Environ. 15, 1653-1657.

15. Rahn, K. A. (1981). The Mn/V ratio as a tracer of large-scale sources of pollution aerosol for the Arctic. Atmos. Environ. 15, 1457-1464.

16. Kahlil, M. A. K. and Rasmussen, R. A. (1983). Gaseous tracers of Arctic haze. Environ. Sci. and Technol. 17, 157-164.

17. Rahn, K., Lowenthal, D., and Lewis, N. (1983). Elemental tracers and source areas of pollution aerosol in Naragansett, R.I., quoted in Air Pollut. Control. Assoc. J. 33, 48.

18. Husain, L., Webber, J., and Canelli, E. (1983). Erasure of midwestern Mn/V signature in an area of high vanadium concentration. APCA Journal 33, 1185-1188.

19. Rasmussen, R. A. (1984). Oregon Graduate Center, private communication.

20. Clarke, J. F., Clark, T. L., Ching, J. K. S., Haagenson, P. L., Husar, R. B., and Patterson, D. E. (1983). Assessment of model simulation of long distance transport. Atmos. Environ. 17, 2449-2462.

21. Zak, B. D. (1981). Lagrangian measurements of sulfur dioxide to sulfate conversion rates. Atmos. Environ. 15, 2583-2591.

22. Zak, B. D. (1984). Sandia National Laboratories, private communication.

23. Clark, T. L., and Clarke, J. F. (1984). A Lagrangian study of the boundary layer transport of pollutants in the northeastern United States. Atmos. Environ. 18, 287-297.

24. Calvert, J. G. (1976). Test of the theory of ozone generation in Los Angeles atmosphere. Environ. Sci. and Technol. 10, 248-256.

25. Alkezweeny, A. J. (1980). Gas to particle conversion in urban plumes. Air pollution Control Association 73rd Annual Meeting, Montreal, Quebec, Canada, June 22-27, paper 80-51.2.

26. Newman, L. (1981). Atmospheric oxidation of sulfur dioxide: A review as viewed from power plant studies and smelter plume studies. Atmos. Environ. 15, 2231-2239.

27.   Wilson, Jr., W. E. (1981). Sulfate formation in point source plumes: A review of recent field studies. Atmos. Environ. 15, 2573-2581.

28.   Levy, A., Drewes, D. R., and Hales, J. M. (1976). SO_2 Oxidation in Plumes: A Review and Assessment of Relevant Mechanistic and Rate Studies. US EPA Report 450/3-76-022. Available NTIS PB 264 206.

29.   Symposium on Plumes and Visibility - Measurements and Model Components, Grand Canyon, AZ, 1980; Atmos. Environ. 15,(10-12) 1981.

30.   Gillani, N. V. (1977). Project MISTT: Mesoscale plume modeling of the dispersion, transformation and ground removal of SO_2. Atmos. Environ. 12, 569-588.

31.   Easter, R. C., Busness, K. M., Hales, J. M., Lee, R. N., Arbuthnot, D. A., Miller, D. F., Sverdrup, G. M., Spicer, C. W., Howes, Jr., J. E., (1980). Plume Conversion rates in the SURE region, Volume I. Electric Power Research Institute Report EA1498, Vol. 1.

32.   Hobbs, P. V., Hegg, D. A., Eltgroth, M. W., and Radke, L. F. (1978). Evolution of particles in the plumes of coal-fired power plants - I. Deductions from field measurements. Atmos. Environ. 12, 935-951.

33.   Sievering, H., Cooke, J., and Pueschel, R. (1981). Importance of deposition velocity for sulfur gas to sulfate particle transformation rates at the Four Corners Power Plant. Atmos. Environ. 15, 2593-2596.

34.   Williams, D. J., Carras, J. N., Milne, J. W., and Heggie, A. C. (1981). The oxidation and long-range transport of sulphur dioxide in a remote region. Atmos. Environ. 15, 2255-2262.

35.   Newman, L., Forrest, J., and Manowitz, B., (1975). The application of an isotopic ratio technique to a study of atmospheric oxidation of sulfur dioxide in the plume from an oil-fired power plant. Atmos. Environ. 9, 959-968.

36.   Forrest, J. and Newman L. (1977). Oxidation of sulfur dioxide in the Sudbury smelter plume. Atmos. Environ. 11, 517-520.

37.   Schwartz, S. E., and Newman, L. (1978). Processes limiting oxidation of sulfur dioxide in stack plumes. Environ. Sci. Technol. 12, 67-73.

38.   Freiberg, J. (1977). Conversion limit and characteristic time of SO_2 oxidation in plumes. Atmos. Environ. 12, 339-347.

39.   Eatough, D. J., Richter, B. E., Eatough, N. L., and Hansen, L. D.
      (1981). Sulfur chemistry in smelter and power plant plumes in the
      western U.S. Atmos. Environ. 15, 2241-2253.

40.   Eatough, D. J., Christensen, J. J., Eatough, N. L., Hill, M. W., Major, T.
      D., Mangelson, N. F., Post, M. E., Ryder, J. F., and Hansen, L. D.
      (1982). Sulfur chemistry in a copper smelter plume. Atmos. Environ.
      16, 1001-1015.

41.   Huang, A. A., Farber, R. J., Mahoney, R. L., Eatough D. J.,
      Hansen, L. D. and Allard, D. W. (1982). Chemistry of invisible power
      plant plumes in southern California - The airborne perspective. Air
      Pollution Control Association, Paper 82-24.5.

42.   Eatough, D. J., Arthur, R. J., Eatough, N. L., Hill, M. W., Mangelson, N.
      F., Richter, B. E., Hansen, L. D., and Cooper, J. A. (1983). SO_2 to
      sulfate conversion rate in an oil-fired power plant plume in a fog bank.
      Air Pollution Control Association 76th Annual Meeting, Atlanta, GA,
      June 19-24, Paper 83-31.1.

43.   Weber, E., (1970). Contribution to the residence time of sulfur dioxide
      in a polluted atmosphere. J. Geophys, Res. 75, 2909-2914.

44.   van Dop, H., Ridder, T. B., den Tonkelaar, J. F., and van Egmond, N. D.
      (1980). Sulfur dioxide measurements on the 213 meter tower at
      Cabauw, the Netherlands. Atmos. Environ. 14, 933-945.

45.   Manowitz, B., Newman, L., and Tucker, W. D., (1972). The Atmospheric
      Diagnostics Program at Bookhaven National Laboratory: Fourth Status
      Report. Brookhaven National Laboratory Report BNL 50361.

46.   Dennis R., Billings., C. E., Record, F. A., Warneck, P., Arin, M. L., and
      Sing, C. Y. (1969). Measurements of sulfur dioxide losses from stack
      plumes. Air Pollution Control Association 62nd Annual Meeting,
      New York, NY, June 26, 1969, Paper 69-156.

47.   Flyger, H., Lewin, E., Lund Thomsen, E., Fenger, J., Lyck, E., and
      Gryning, S. E. (1978). Airborne investigations of SO_2 oxidation in the
      plumes from power stations. Atmos. Environ. 12, 295-296.

48.   Flyger, H. and Fenger, J., (1976). Conversion of sulfur dioxide in the
      atmosphere. Z. Anal. Chem. 282, 297-300.

49.   Alkezweeny, A. J. and Powell, D. C. (1977). Estimation of
      transformation rate of SO_2 to SO_4 from atmospheric concentration
      data. Atmos. Environ. 11, 179-182.

50. Forrest, J., Schwartz, S. E., and Newman, L. (1978). Conversion of sulfur dioxide to sulfate during the da Vinci flights. Atmos. Environ. 13, 157-167.

51. Chang, T. Y. (1979). Estimate of the conversion rate of SO_2 to SO_4 from the da Vinci flight data. Atmos. Environ. 13, 1663-1664.

52. Lewis, C. W. and Stevens, R. K. (1984). Hybrid receptor model for secondary sulfate from a coal-fired power plant, Submitted to Atmos. Environ.

53. White, W. H. (1977). NO_x-O_3 photochemistry in power plant plumes: comparison of theory with observation. Environ. Sci. Technol. 11, 995-1000.

54. Miller, D. F., Alkezweeny, A. J., Hales, J. M., and Lee, R. N. (1978). Ozone formation related to power plant emissions. Science 202, 1186-1188.

55. Calvert, J. G. (1976). Hydrocarbon involvement in photochemical smog formation in Los Angeles atmosphere. Environ. Sci. Technol. 10, 256-262.

56. Spicer, C. W., (1982). Nitrogen oxide reactions in the urban plume of Boston. Science, 215, 1095-1097.

57. Chang, T. Y., Norbeck, J. M., and Weinstock, B. (1979). An Estimate of the NO_x removal rate in an urban atmosphere. Environ. Sci. Technology. 13, 1534-1537.

58. Spicer, C. W. (1977). The fate of nitrogen oxides in the atmosphere, in Adv. Environ. Sci. Technol. 7, J. N. Pitts and R. L. Metcalf, Eds., Wiley, New York, pp. 163-261.

59. Anderson, L. G. (1983). Fate of nitrogen oxides in urban atmospheres, in Advan. Environ. Sci. Technol. 12 S. E. Schwartz, Ed., Wiley, New York, pp. 371-409.

60. Roberts, J. N., Fehsenfeld, F. C., Liu, S. C., Bollinger, M. J., Hahn, C., Albritton, D. L., and Sievers, R. E., (1983). Measurements of aromatic hydrocarbon ratios and NO_x concentrations in the rural troposphere: estimation of air mass photochemical age and NO_x removal rate. Atmos. Environ., in press.

61. Fehsenfeld, F. C., Bollinger, M. J., Liu, S. C., Parrish, D. D.,
     McFarland, M., Trainer, M., Kley, D., Murphy, P. C., Albritton, D. L.,
     and Lenschow, D. H. (1983). A study of ozone in the Colorado
     mountains. J. Atmos. Chem. 1, 87-105.

62. Nguyen, B. C., Bonsang, B., Lambert, G., and Pasquier, J. L. (1975).
     Residence time of sulfur dioxide in the marine atmosphere. Pure and
     Applied Geophys. 113, 489-500.

63. Nonlinearity of Acid Precipitation Processes--Preliminary Program
     Plan (1983). Brookhaven National Laboratory, Report BNL 33164.

64. Gatz, D. F., (1977). Review of chemical-tracer experiments on
     precipitation systems, in Precipitation Scavenging (1974),
     R. G. Semonin and R. W. Beadle, coordinators, Technical Information
     Center, ERDA, CONF-741003, pp. 321-336; Atmos. Environ. 11, 945-
     953.

65. Davis W. E. and Young, J. A. (1977). Results of in-cloud tracer releases
     in frontal storms in Precipitation Scavenging (1974), R. G. Semonin and
     R. W. Beadle, coordinators, Technical Information Center, ERDA,
     CONF-741003, pp. 337-361.

66. Young, J. A., Tanner, T. M., Thomas, C. W., and Wogman, N. A. (1977).
     Entrainment of tracers into the convective clouds at 10,000 to
     13,500 feet near St. Louis, in Precipitation Scavenging (1974), R. G.
     Semonin and R. W., Beadle, coordinators, Technical Information Center,
     ERDA, CONF-741003, pp. 362-381.

67. Dingle, A. N. (1977). Scavenging and dispersal of tracer by a self-
     propagating convective shower system, in Precipitation Scavenging
     (1974), R. G. Semonin and R. W. Beadle, coordinators, Technical
     Information Center, ERDA, CONF-741003, pp. 395-424.

68. Schwartz, S. E., (1984). Gas-aqueous reactions of sulfur and nitrogen
     oxides in liquid water clouds. In SO_2 and NO_2
     Oxidation Mechanisms: Atmospheric Considerations, J. G. Calvert, ed.,
     Butterworth, Boston, pp. 173-208.

69. Senum, G. I., (1984). Brookhaven National Laboratory, personal
     communication.

70. Altshuller, A. P. (1983). Review: natural volatile organic substances
     and their effect on air quality in the United States. Atmos. Environ.
     17, 2131-2165.

71.   Flyger, H., Lewin, E., Lund Thomsen, E., Fenger, J., Lyck, E., and Gryning, S. E. (1977). Physical and Chemical Processes of Sulfur Dioxide in the Plume from an Oil-Fired Power Station. Research Establishment Ris?, Copenhagen, Denmark, Report No. 328.

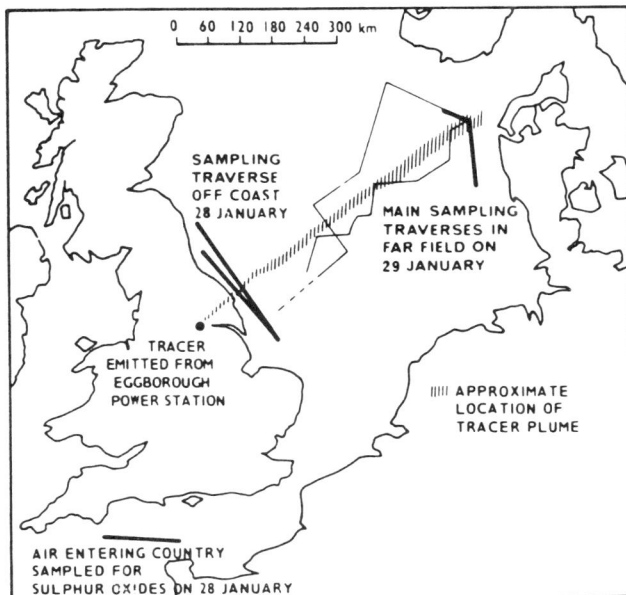

Fig. 1.   Flight path and main sampling traverses on 28 and 29 January
1981. (Ref. 10)

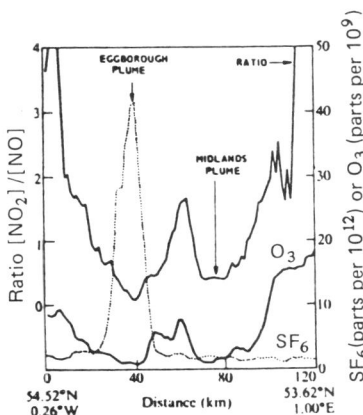

Fig. 2.   Cross-plume profile,
28 January 1981. Height
150 m. Distance from
Eggborough Power Station,
105 km. (Ref. 9)

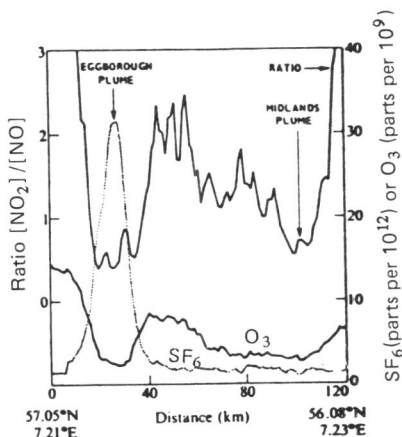

Fig. 3.   Cross-plume profile.
29 January 1981.
Height 150 m. Dis-
tance from Eggborough
Power Station, 650 km.
(Ref. 9)

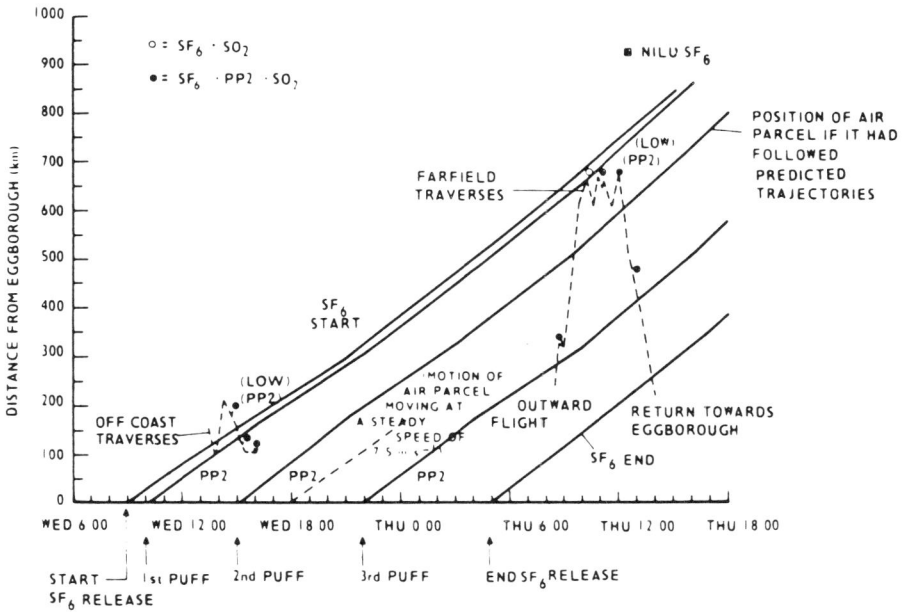

Fig. 4.   Distance-time plot for aircraft and air parcel movement on 28
          and 29 January 1981. (Ref. 10)

RAMPED TRACER RELEASE

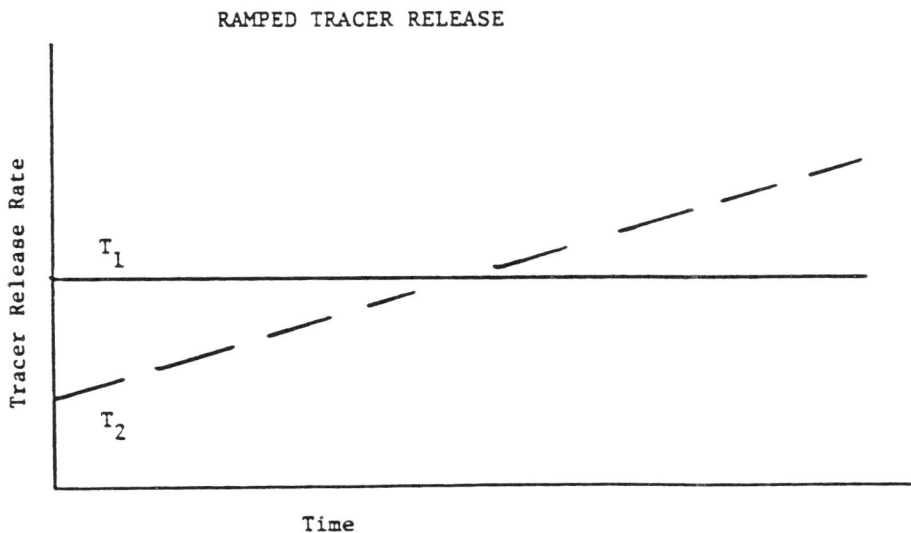

Fig. 5.   Suggested means of obtaining Lagrangian marker with continuous
release of two tracers. Tracer 1 is released at constant rate;
release rate of tracer 2 is increased with time. Age of sample may
be evaluated from ratio of two tracer concentrations.

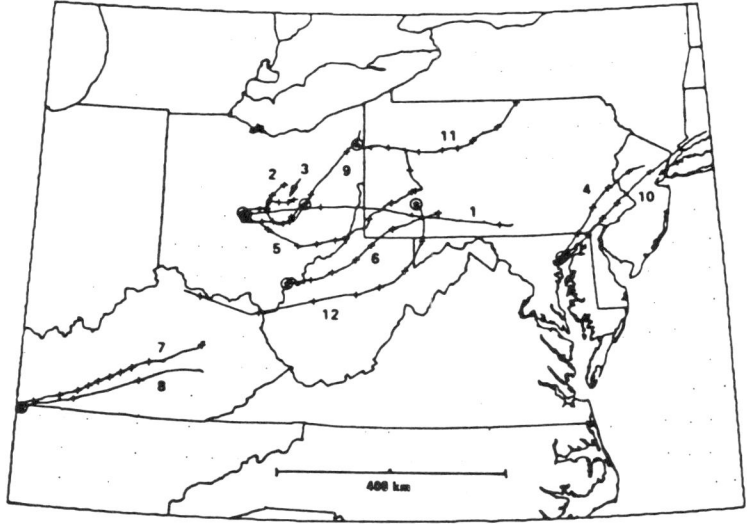

Fig. 6.   Tracks of EPA tetroons with 3-h travel segments separated
           by (+). (Ref. 20)

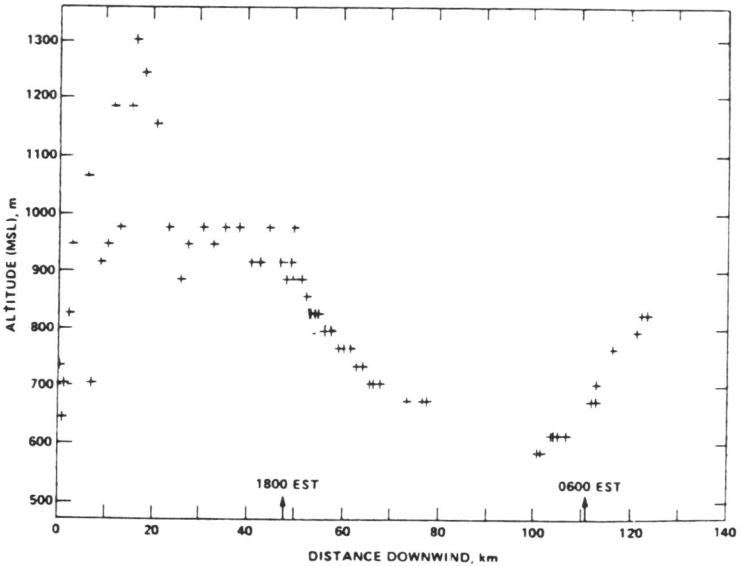

Fig. 7.   Plots of tetroon height vs distance downwind for EPA
           tetroon flight 3 released from Columbus, OH, 1545 GMT on 25
           July 1980. (Ref. 20)

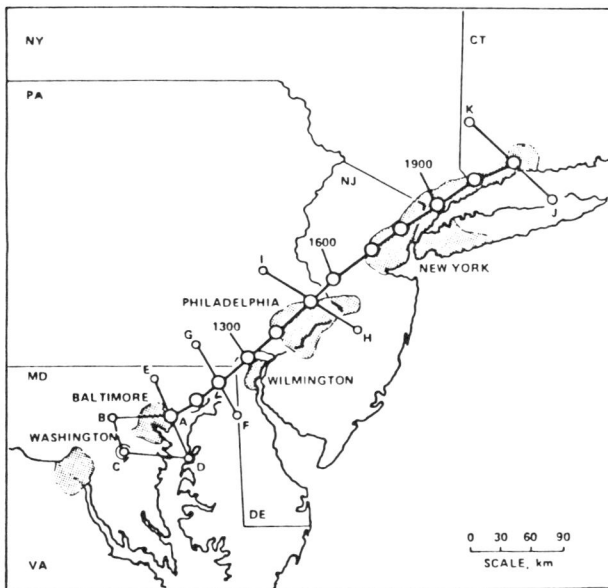

Fig. 8.   The hourly tetroon positions (larger circles), the upwind
           sampling flight path (A-B-C-D), and the four Lagrangian sampling
           transect series (D-E,F-G,H-I, and J-K) in the northeastern United
           States on 14 August 1980. Martin State Airport is located at
           point A. (Ref. 23)

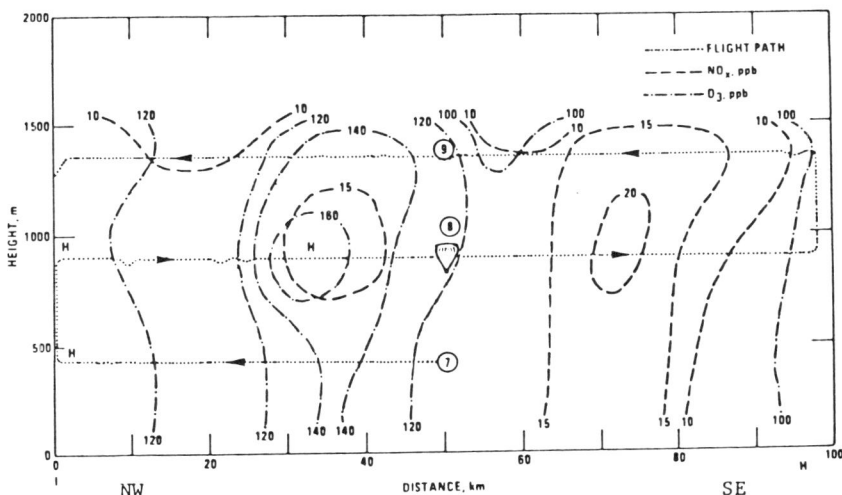

Fig. 9.   The NO_x and O_3 cross sectional analysis of the Baltimore-
           Washington and Philadelphia urban plumes along sampling transect
           series H-I, 160 km downwind of downtown Baltimore, from 1534
           to 1641 EST. (Ref. 23)

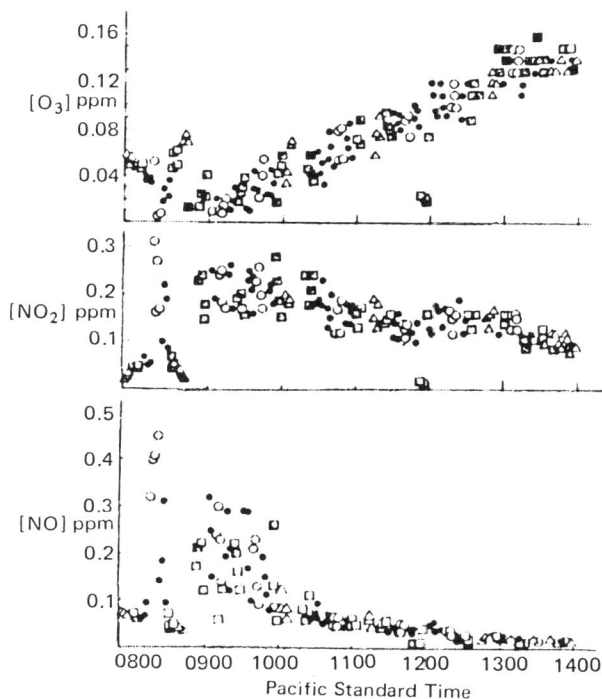

Fig. 10.   Time dependence of concentrations of NO, NO$_2$, and O$_3$ as monitored in LARPP operation #33. (Ref. 24)

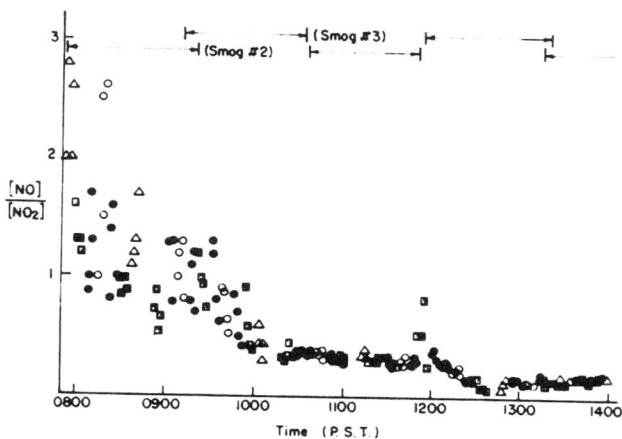

Fig. 11.   ]NO[/]NO$_2$[ratio as a function of time in LARPP operation #33. (Ref. 24)

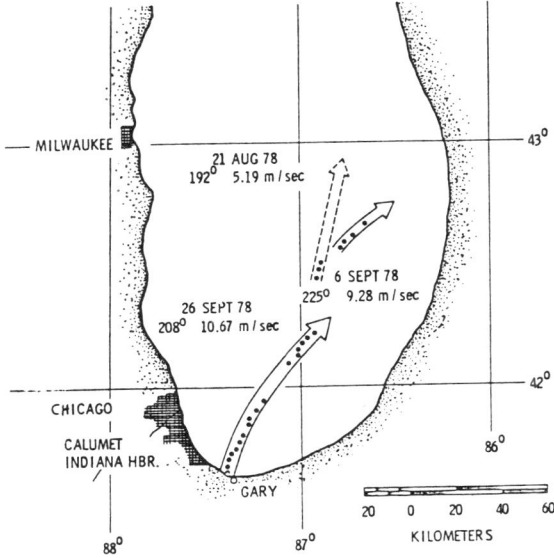

Fig. 12.   Aircraft sampling routes during the August and September
1978 experiments. The arrows show the tetroon trajectories.
Next to the arrows are the wind speed and direction. (Ref. 25)

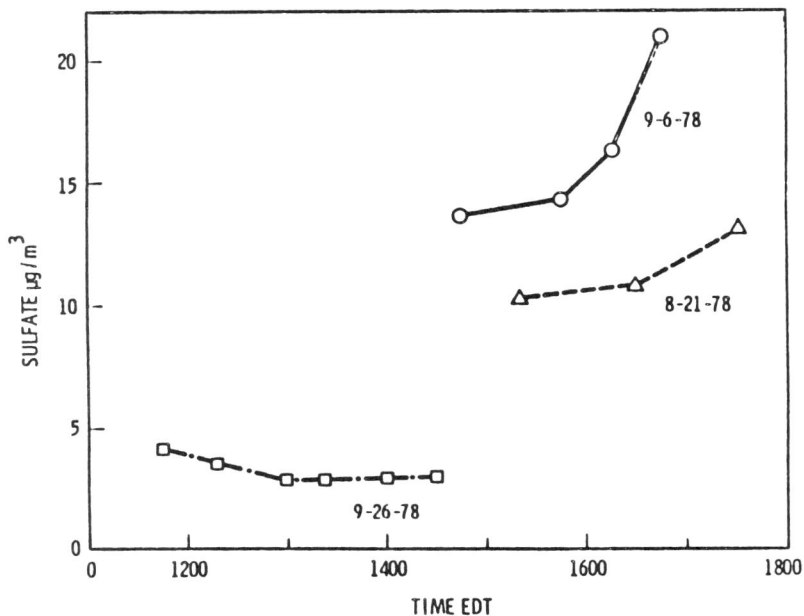

Fig. 13.   Sulfate concentration downwind of Gary, IN, August 21 and
           September 26, 1978, and downwind of Chicago, IL, on September
           6, 1978. (Ref. 25)

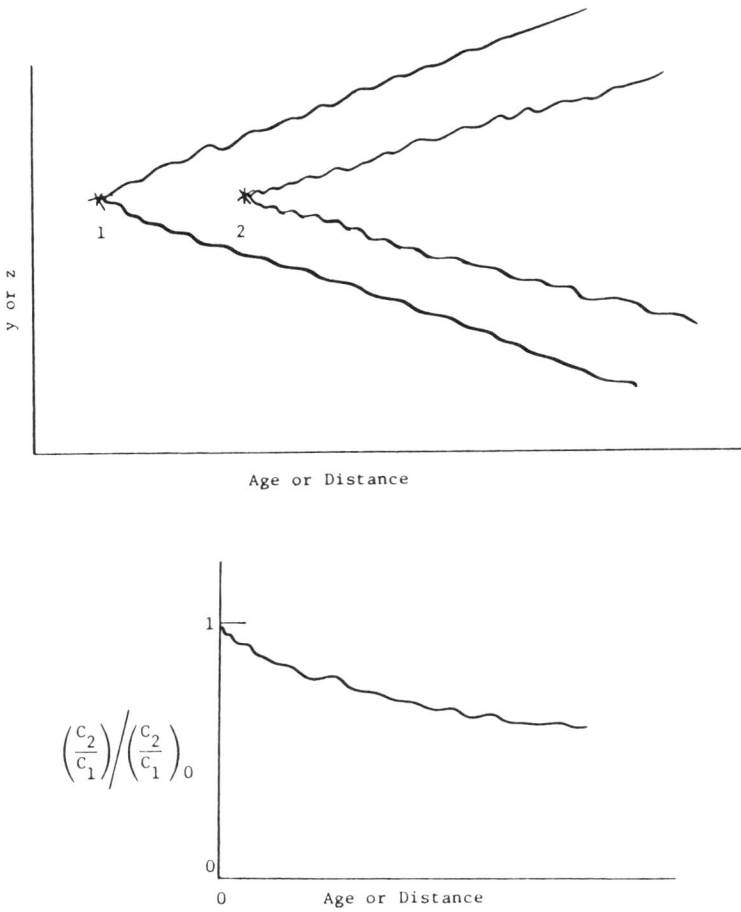

Fig. 14. Schematic illustration of error in ratioing concentrations of species
not emitted from the same source. (a) Sketch of plume dispersal
from two effective sources (1) and (2). (b) Ratio of concentrations
of conservative tracers emitted at points (1) and (2).

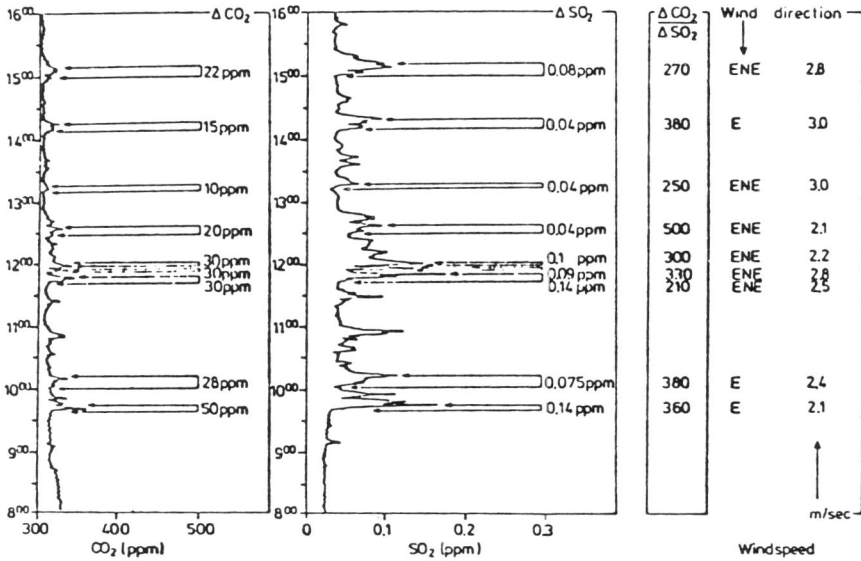

Fig. 15.   Copy of a part of the recorder traces of the continuous recording of the CO_2 and the SO_2 concentrations at a sampling site in Frankfurt am Main in the center of the city on August 22, 1967. The ordinate is the time (LT) and the abscissa the concentration in ppm. The values ΔCO_2 and ΔSO_2 are the differences in concentration between the beginning and the maximum of parallel increases. (Ref. 43)

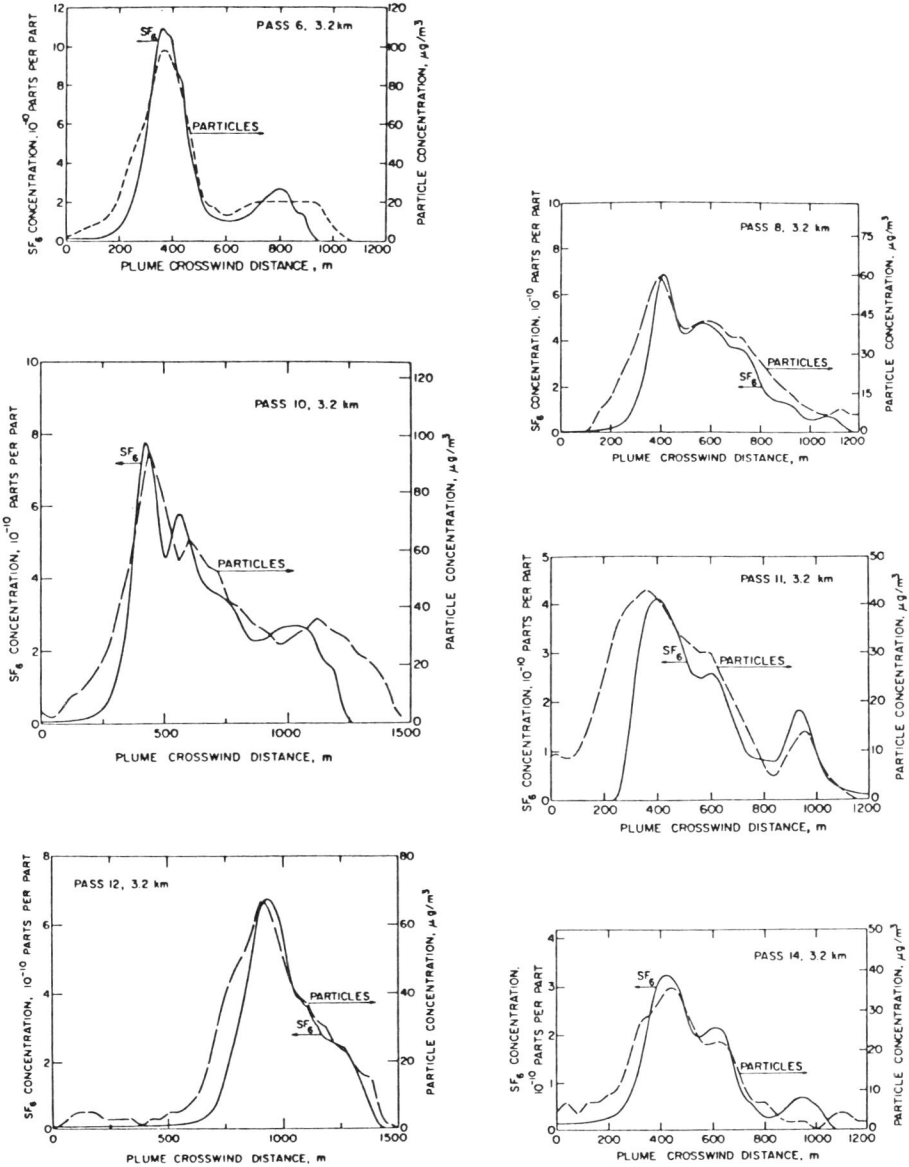

Fig. 16.   Continuous recorder tracers of SF_6 and b_{scat} in aircraft cross-plume traverses of coal-fired power plants. (Ref. 45)

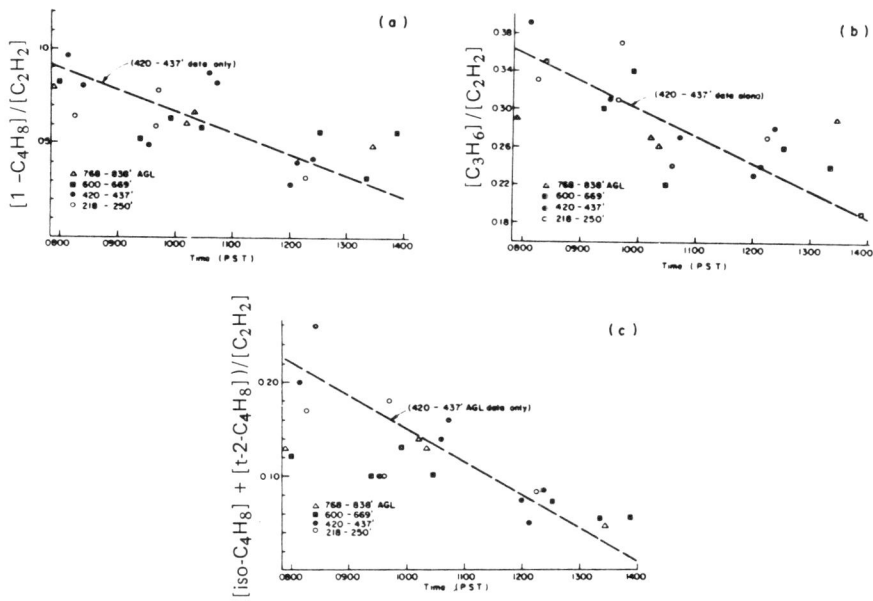

Fig. 17.   Ratios of concentrations of reactive hydrocarbons to C_2H_2 in IARPP operation 33, 5 November, 1973. a, $1\text{-}C_4H_8$. b, C_3H_8. c. iso-C_4H_8 + t-$2C_4H_8$. (Ref. 55)

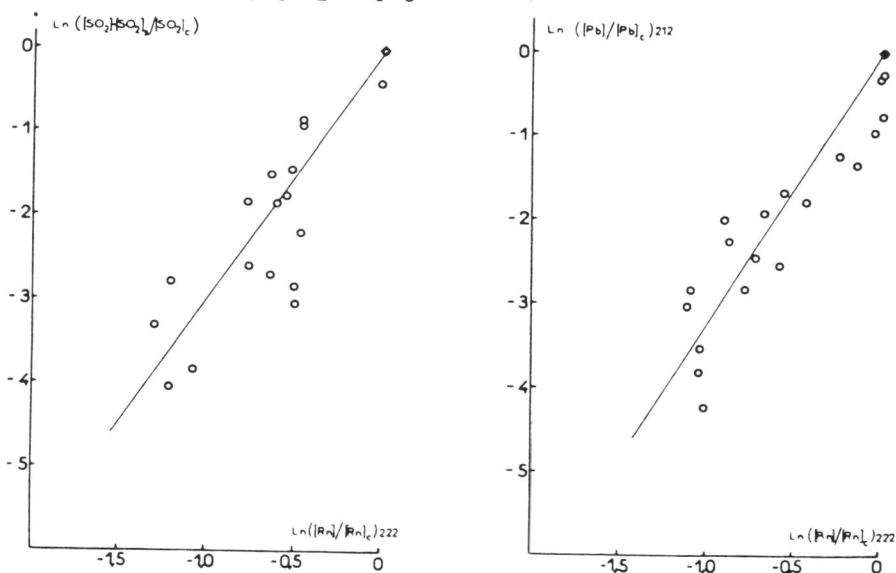

Fig. 18.   Plot of relative concentrations SO_2 and cloud tracers. Subscript c denotes continental concentrations. Measurements were taken along coast and southwards into Mediterranean Sea, 19-26 April 1974. (Ref. 62)

TRACER APPLICATIONS FOR THE STUDY OF PRECIPITATION PROCESSES

Richard G. Semonin

Illinois State Water Survey
Champaign, IL 61820

I. INTRODUCTION

Before embarking on the use of tracers to improve our knowledge of precipitation processes, I believe it is essential to touch on some of the unknowns about these processes and then pose some thoughts as to how tracers might be used to improve our knowledge. At the risk of oversimplifying the precipitation process, we should remember that there are three necessary and sufficient conditions for precipitation. The first is an adequate supply of water vapor to the process; second, an appropriate supply of either cloud condensation or ice-forming nuclei or both; and third, the presence of appropriate atmospheric motions conducive to the interaction of the first and second in the formation of either liquid or solid cloud particles, the precursors to precipitation.

II. WATER VAPOR AVAILABILITY

Viewing the precipitation reaching the ground as a condensed form of a gas, one is left to question the source or sources of the original water vapor later observed as either liquid or solid. We know that water vapor is a minor consituent of the atmosphere and is highly variable in both space and time. We further know it enters the atmosphere in its lowest layers by evaporation from oceans, lakes, and land surfaces and by evapotranspiration from vegetation. The flux of water vapor from the earth's surface is seasonally and latitudinally dependent. The return of the vapor in the form of precipitation, frost, and dew completes a major part of the general hydrologic cycle. Precipitation is the primary removal process of atmospheric water vapor; frost and dew are only minor contributors.

Obvious to the experience of many, precipitation is zero most of the time with very large values occurring in relatively short time periods at a point. It is this very nature of precipitation, its relatively infrequent occurrence, that is one of the obstacles to progress toward greater knowledge of precipitation processes.

Equally obvious, the upward flux of water vapor and the downward flux of precipitation must be equal over the globe when measured over an appropriate time scale, but because of large local variability, the water vapor transport remains very complex on all motion scales including the general circulation of the atmosphere. The transport of water vapor is important even in times of no precipitation. It has been estimated that the daily horizontal transport of water vapor across the state of Illinois, expressed as grams of available precipitation, is 7.6×10^{15} g[1]. Of this immense value, only 5% reaches the surface as precipitation. Although this precentage is small, the amount of water is still tremendous --3.8×10^{14}g. These numbers are given to impress all investigators with the scope of the problem and the relative inefficiency of the atmosphere in processing the available water vapor. Whereas to some this assessment may appear of trival interest, it must also be remembered that because of its high latent heat of condensation, water vapor is important to the transport of energy. The release of the latent energy in cyclonic as well as localized convective storms is a major factor in developing the character and intensity of precipitation.

The application of information regarding source apportionment of water vapor relates to: (1) differentiating between the role of the Pacific Ocean, Gulf of Mexico, Atlantic Ocean, and various local sources for the development of precipitation. This problem is analogous to following a pollutant from a source through transport and transformation processes to its deposition. To achieve a higher level of knowledge for these equivalent water vapor processes, it is first advantageous to be knowledgeable about the source of relative contributions from multiple sources.

An example of the diversity of the source term for water vapor in the upper Midwest and Great Lakes is shown in Fig. 1 (Ref. 3). The most striking feature in Fig. 1 is the very dry air through the lower Mississippi valley, which is associated with a strong anticyclone over the central US. This pattern was the major contributing factor to the drought of the mid-1930s. However, in the midst of this drought, the upper Great Lakes states received above normal precipitation, as is shown by the shading in the small inset map in Fig. 1. It is somewhat obvious that the water vapor source for this above normal precipitation is the Pacific Ocean feeding across the mountain barriers of Mexico and the US. This case study is a very good example of

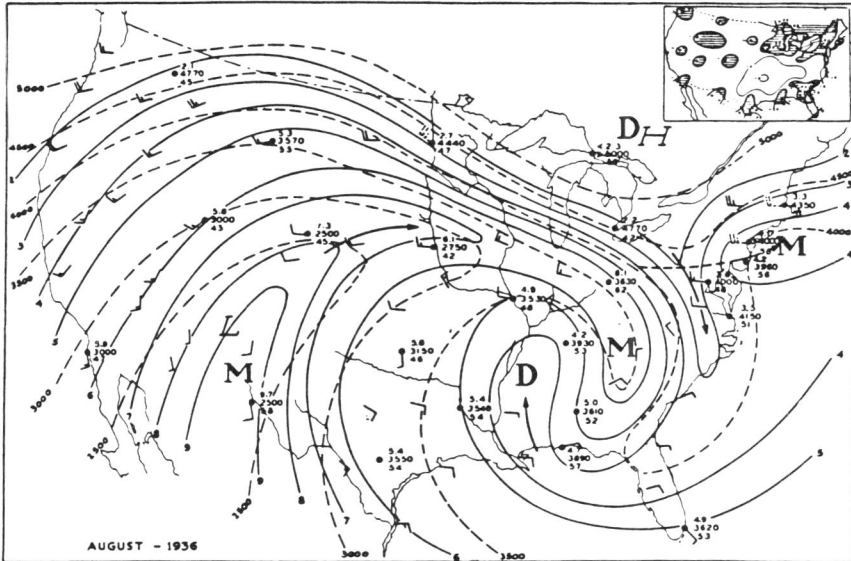

Fig. 1.   Isentropic chart for the 315°A surface for August 1936. Solid
lines are isopleths of mixing ratio in grams per kilogram; broken
lines represent (in meters) the height of the isentropic surface;
moist and dry tongues are labeled M and D, respectively. The
insert showsdeparture from normal precipitation for the month (in
inches); shaded areas show excess, and unshaded areas show
deficit (Ref. 3).

source knowledge, but the usual precipitation systems are much more difficult to evaluate. Clearly, there is a need to identify and separate the long-distance sources as opposed to the contribution from local evaporation and vegetative transpiration.

III. WATER AND ICE NUCLEI

The combination of water vapor and the presence of nuclei are necessary conditions for the formation of clouds--the precursors to precipitation. The upward movement of air in the free atmosphere, which causes expansion and cooling and transports vapor and nuclei, is the final condition required for cloud formation and will be discussed in the next section.

The process of condensation is not one that occurs readily in an environment free of impurities. Some of the earliest laboratory experiments confirmed that in the absence of nuclei, super saturation of 400% may be necessary before condensed droplets become visible by light scattering.[2] Fortunately, the real atmosphere contains numerous particles, crystals, and even solution droplets that stand ready to serve as condensation centers at very low supersaturations. The supersaturation during cloud initiation ranges between a few hundreths of a per cent up to perhaps a few tens of per cent. Obviously, there is competition for the available supply of water vapor between the various nuclei. Those that require less supersaturation remove some water vapor, reducing the supersaturation and decreasing the opportunity for nuclei that require greater supersaturation to become active in the process.

The chemical and physical properties of the nuclei determine their effectiveness in forming water droplets of ice crystals. In relatively pure form, the effectiveness of various nuclei to form water or ice particles can be determined. However, the atmospheric aerosol is thought to arise from three primary ways, none of which will yield chemically homogeneous nuclei. These three means of nuclei production are (1) the condensation and sublimation of vapors in the formation of smokes and gaseous reactions, (2) the formation of dust and spray by mechanical disruption and dispersion of matter, and (3) the coagulation of existing nuclei, leading to larger particles of even greater mixed constitution.[4] The inhomogeneity of the atmospheric aerosol results in great difficulty identifying individual sources.

A generalized picture of the atmospheric aerosol size distribution is shown in Fig. 2. In this figure, the size of the aerosol extends from nanometers in radius for small ions to micrometers for giant nuclei, that is, a range of greater than 4 orders of magnitude. However, as mentioned previously, the competition for water vapor precludes activation of most of the aerosol. Cloud condensation nuclei (CCN) are usually considered as that portion of the size distribution (shown in Fig. 2) classified as large and giant particles. The chemical speciation of the fraction of the CCN population that become cloud particles is not documented well and is only alluded to by a qualitative judgment of the source in notes 8, 9, and 10 of Fig. 2.

An additional complexity is the fact that an evaporated cloud "releases" used nuclei into the atmosphere for a repetition of the cloud process. The number of times that an original nuclei particle undergoes condensation and evaporation prior to its deposition by precipitation is largely unknown, although it has been estimated as greater than 10 times.[6] Each time the "wet" process occurs and is followed by evaporation, the microphysical cloud processes (including droplet or crystal pair interactions) lead to a chemical reformulation of the evaporated aerosol.

Because the period of time between the appearance of a nucleus to its deposition in precipitation may be many hours, and perhaps even days, the task of identifying particular sources from the viewpoint of a particular receptor becomes immense. The complexities of this problem are underlined by the various factors touched on above, and the result is the development of great inhomogeneities of nuclei chemistry during transport and transformation.

IV. ATMOSPHERIC MOTIONS

The final, and sufficient, condition for the formation of clouds and eventually precipitation is the appropriate movement of the water vapor and nuclei in the free atmosphere. A parcel of air rising in the atmosphere undergoes cooling and a reduction in the partial pressure of water vapor. Although the dewpoint temperature cools in an ascending parcel, it does so at a slower rate than the temperature. Eventually the temperature overtakes the dewpoint in a continuously rising parcel and condensation begins. A further contribution to the upward speed arises because of the release of latent heat to the parcel during the change of phase of the water vapor.

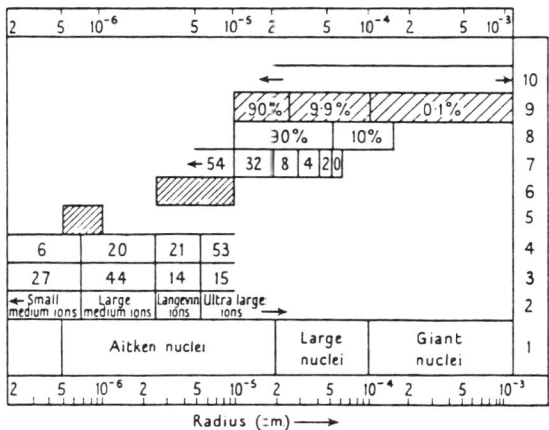

Fig. 2.   The size range of atmospheric condensation nuclei taken from many sources. For details see Ref. 5. The meaning of the numbers along the ordinate are as follows:

1. Classification of nuclei into size groups used in Ref. 1.
2. Classification of atmospheric ions.
3. Per cent distribution of total ions in country air.
4. Per cent distribution of total ions at Frankfort-on-Main.
5. Size range of newly formed gas-flame ions.
6. Size range of Aitken nuclei by electromicrographs.
7. Per cent distribution of haze droplets in country air.
8. Size distribution of atmospheric dust.
9. Size distribution of continental aerosol at Frankfort.
10. Size distribution of sea-salt nuclei over oceans.

The atmospheric motions important to the cloud formation process can be categorized into four compartments. These four categories of motion are (1) the relatively slow movement of air constrained to gently sloping isentropic surfaces, including certain fronts; (2) the mechanically induced motion caused by terrain barriers; (3) convective motion, largely in response to differential surface heating but also influenced by sharply defined fronts; and (4) the turbulence resulting from ground friction, resulting in random stirring of the atmosphere below an inversion.[7] The first three of these motions influence the formation of clouds and precipitaton profoundly while the relative role of the fourth category is little understood at this time.

At first glance, these categories of motion appear independent, whereas, in fact they are dependent upon one another to varying degrees, adding further to our incomprehension of the precipitation process. The clouds associated with upglide motion on fronts and other baroclinic disturbances are similar to the frequently observed clouds formed by mountain barriers, and both are occasionally associated with embedded convective elements. The stratiform clouds at mid- and upper-levels of the atmosphere directly resulting from strong convection further exemplify the less than independent character of the motion categories.

The relatively gentle upglide motion is usually correlated to specific patterns of large-scale convergence, which in turn, predisposes the atmosphere to organized convecton on the mesoscale. Similarly, organized convection systems can modify the larger scale flow in ways not yet fully appreciated.

Figures 3 and 4 serve as illustrations of two concepts of motion associated with convective storms. In Fig. 3 (Ref. 8), a schematic of the motions in the environment and inside a convective storm are shown in simple form. Even this simple picture was used to explain the growth of flanking clouds and thus propagating the storm.[8] A more complicated model of air motion in proximity to a front is shown in Fig. 4 (Ref. 9). Both of these models show the mutual interaction between the atmospheric environment and the storm. Figure 4 was not constructed from measurements but represents a concept based on the surface-measured variables of pressure, temperature, humidity, wind speed, and wind direction. The verticality of the picture is inferred from remote measures of radar reflectivity and a scattering of aircraft probes. What is needed is a means of validating this model as well as other forms of convective cloud systems. Knowledge of the partitioning between these motions and the role of turbulence in the precipitation process remain fruitful research topics.

V. PRECIPITATION FORMATION AND DURATION

The two primary forms of precipitation, rain and snow, are to many people the most common form of experienced weather and are the final product of the coexistence of the three conditions briefly described in the foregoing material. Soon after these conditions are satisfied and a size spectrum of liquid or solid water particles is formed, microphysical processes begin to deform the initial spectrum by continued condensation and aggregation.

To illustrate this point, consider the information shown in Table 1 (Ref. 4). The nucleation process varies markedly with the time required to grow a 5-μm-radius droplet extending from as low as 30 s to as high as 2,450 s. There are several points implicitly expressed in this table, but two are immediately noteworthy. This first is the general indication of a decrease in the time of appearance of 5-μm droplets with increasing parcel rate of ascent. Second, the spread within a fixed ascent-rate column indicates the efficiency of the CNC-size spectrum for producing 5-μm droplets.

Another way of viewing this problem is shown in Fig. 5 (Ref. 10). These calculations assumed the total aerosol was composed of soluble ammonium sulfate and other insoluble material. The size distribution was prescribed, but various percentages from 0 to 100 of ammonium sulfate were considered. The size distribution of the simulated cloud was calculated and developed as a result of condensation and collection. A uniform ascent rate was imposed, but the supersaturation varied in response to the condensation process. The calculation was terminated when a minimum detectable radar signal (MDS) level was achieved. The time required to reach the MDS ranged from 1,410 s to 1,680 s for the 0 to 100% soluble CCN spectrum, respectively.

At this point, let's return to the concepts of motion leading to cloud formation very briefly and focus on only the relatively slow upglide motion and the more vigorous convective motion producing stratiform and cumuliform clouds. In the former, the sustaining upward motion would be

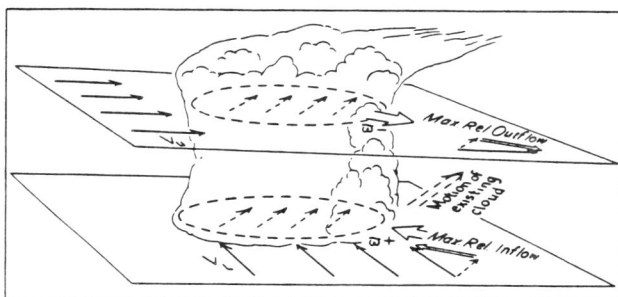

Fig. 3.   Ambient and in-cloud winds associated with a model convective
           rainstorm imbedded in a shearing current. The model suggests
           relative inflow and outflow on the downshear side, giving rise to
           new turret development. (After Ref. 8).

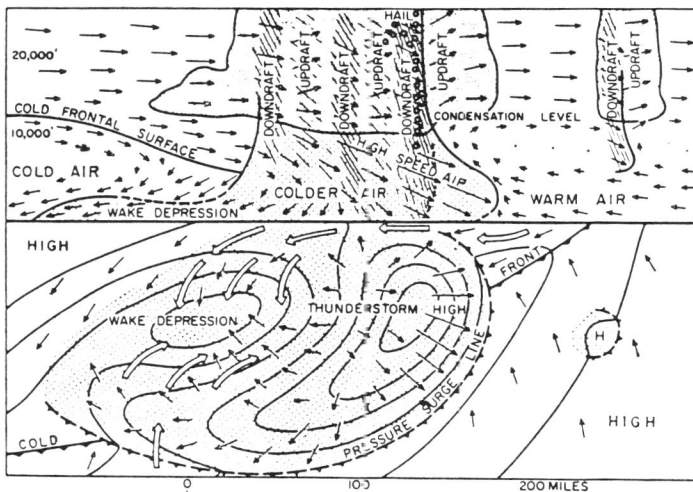

Fig. 4.   A thunderstorm model illustrating the complex interaction
between the ambient flow field and the local perturbations
caused by storm dynamics (Ref. 9). Wind vectors are indicated
by small arrows, and upper motion relative to the moving system
is shown by the large, open arrows on the surface map.

TABLE I. THE GROWTH OF NONUNIFORM GROUPS OF NUCLEI UNDER VARIOUS ATMOSPHERIC CONDITIONS (REF. 4).

Rate of ascent (cm sec^{-1})	400	30	1 x 5
Total nucleus concentration (per cm^3)	500	2,000	667
Range of nuclear masses (moles)	10^{-19} to 10^{-15}	10^{-16} to $10^{-13} \times 8$	10^{-16} to $10^{-13} \times 5$
Peak supersaturation (%)	0 x 36	0 x 072	0 x 056
Time to attain peak supersaturation (s)	25	70	400
Mass range of activated nuclei (moles)	10^{-17} to 10^{-15}	$10^{-15} \times 65$ to $10^{-13} \times 8$	$10^{-15} \times 5$ to $10^{-13} \times 5$
Concentration of activated nuclei (per cm^3)	250	410	40
Radius at peak supersaturation (μ)	1 x 5 to 3	1 x 5 to 4 x 5	2 to 8
Time taken to grow to r = 5 μ (s)	150 to 100	2,200 to 100	2,450 to 30
Modulus of size distribution[a] (m) at 100 m above cloud base	0 x 06	0 x 33	0 x 23
Median-volume radius at 100 m above cloud base	~3 μ	~5 x 5 μ	~14 μ

[a] $m = \sqrt{(2\sigma/r_n)}$ where r_n = median-volume radius and σ = standard deviation of drop volumes. A small value of m denotes a more nearly uniform spectrum.

measured in centimeters per second, whereas the latter upward speeds are measured in meters per second. The horizontal motion of a cloudy air parcel in a stratiform cloud is typically the same as the wind and is determined primarily by the pressure gradient at cloud level. The movement of the cloudy element of a cumuliform cloud is influenced considerably by the strong upward speed in addition ot the loud environment horizontal wind.

These dynamic differences between the two major cloud types lead to differences in time scales for the microphysical processes leading to differences in the release of precipitation. Consideration of these concepts is cause to admit to the relative chemical differences expected between clouds and precipitation embedded in the horizontal long-range transport layers (stratiform clouds) and those that penetrate vertically through such layers. The vertical transport of boundary layer pollutants into stratiform clouds is a slow process and will draw upon many sources as the cloud system is advected over great distances. In contrast, cumuliform clouds quickly inhale boundary layer air and subject that air to the in-cloud microphysics very rapidly.

No two cyclonic systems are identical, but all are characterized by associated cold and warm fronts in the middle latitudes. The relatively widespread and slow vertical speeds are schematically inferred from Fig. 6 (Ref. 11). This figure also illustrates the difficulties in following the general atmospheric motion to and through the clouds and precipitation associated with such an extratropical cyclone. The gentle slope of the warm frontal surface gives rise to upward speeds of centimeters per second and the subsiding atmosphere in the cold air mass provides downward speeds of the same magnitude. Of course, in the cloud and precipitating regions, these speeds are locally increased as determined by the system thermodynamics and microphysics. The smooth streamlines depicted in Fig. 6 indicate the general horizontal motion associated w th such a large system; whether any property of an air parcel survives the long-range transport along such a streamline while subjected to gentle and severe upward and downward motions is in need of experimentation.

The growth of cloud particles leading to precipitation development is accomplished by either prolonged condensation (deposition) of vapor to previously nucleated droplets (ice crystals) or by the aggregation of water or

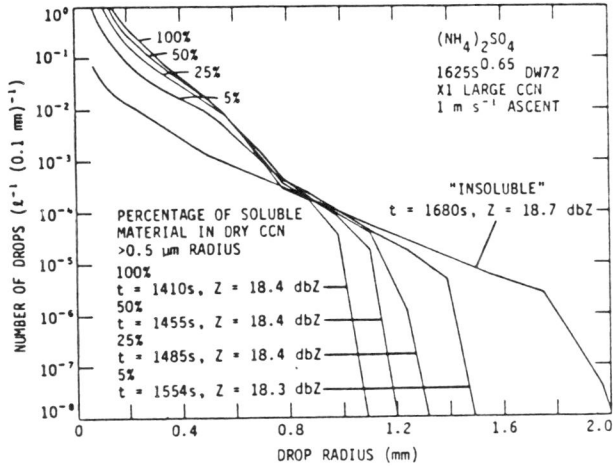

Fig. 5.   Computed raindrop size distributions at a constant radar
reflectivity value for initial CCN properties ranging from totally
insoluble to 100% soluble ammonium sulfate (Ref. 10).

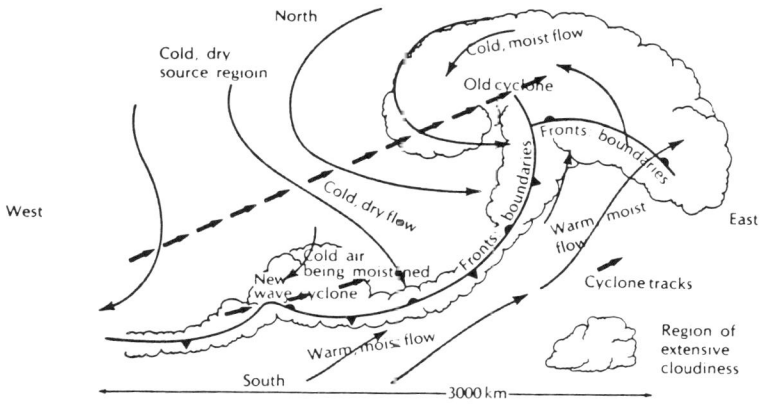

Fig. 6.   A schematic showing the general relationship between air masses, fronts, relative circulation, and cloudiness. Note that the scale of such a system is nearly continental. (Ref. 11).

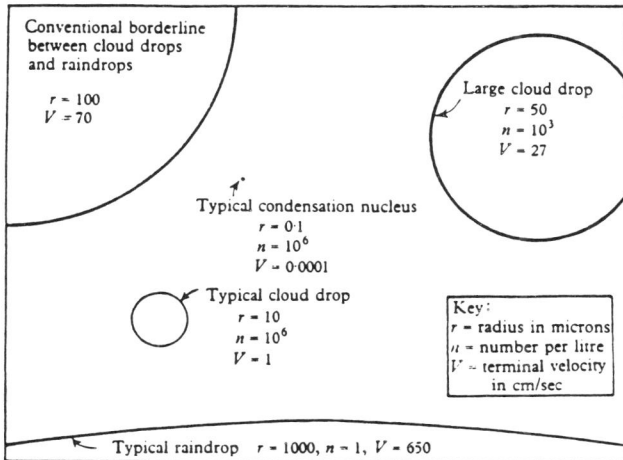

Fig. 7.   The relative size of particles involved in the development of clouds and precipitation (Ref. 12). Note the disparity between the size of a typical cloud droplet (bearing the initial chemistry) and the typical raindrop (the contributor to wet deposition). The 10^6 difference in volume suggests that the rain is an aerosol. In addition, the long acknowledged, but still little understood role of entrainment in altering the cloud dynamics and chemistry must be quantified as a contribution to the total picture of the precipitation process.

ice particles. The rate at which these two broad mechanisms proceed is greatly influenced by the dynamics of the cloud system, the available vapor supply, and the chemical makeup of the cloud nuclei size spectrum. The complexities of the system preclude quantitative predictions of either the quality or quantity of precipitation deposited from a specific storm event. The relative sizes of the particles comprising a precipitating cloud are shown in Fig 7 (Ref. 12). The sizes of particles range over 4 orders of magnitude whereas the number of concentrations extend over 6. From a cloud chemistry point of view, the interaction between these particles and interstitial gases is not well described. The drops may undergo evaporation that alters their original chemical character, they may also absorb gases, and they may serve as a solvent.

VI. TRACER NEEDS

The development and application of suitable tracers must be considered carefully in attempting to address the development of precipitation. The many intermediate, and usually isolated, steps involved in cloud and precipitation initiation involve gases and particles and are not unlike the steps taken to study atmospheric pollution.

As stated earlier, it is clear that tracers are needed to identify water vapor sources. At the very least, a means of partitioning observed water vapor into either oceanic or continental sources would be very helpful. The source identification can be used to infer the transport and dispersion that occurred before the observed deposition. If the tracer is sufficiently unique, it may also lead to clarification of some of the questions embedded in our crude image of precipitating storm dynamics.

A cloud system can be conceptually thought of as a filter operating on both particles and gases. Certain amounts of the particles and gases pass through the filter, depending on the particle size spectrum, their chemical composition, and the absorption-desorption properties of pollutant gases. The cloud system by its very nature is a powerful transport vehicle for pollutants into upper atmospheric layers and not all material is scavenged in the process. Anyone who has observed the anvil of a thunderstorm can readily appreciate that the cirrus cloud represents the high-level injection and transport of water vapor that originates mostly in the subcloud layer. To place the importance of cloud systems, especially convective storms, in the context of a vertical transport medium, some relative scales are shown in Fig. 8 (Ref. 11). Even this picture is not quite accurate as it is now thought that buoyant elements of large thunderstorms may penetrate the tropopause and release material into the stratosphere.

The tracer simulation of both cloud condensation and ice nuclei will add the final technique for quantifying the precipitation processes. A note of caution must be expressed, however, because the entire field of weather modification has been intentionally placing a "tracer" into clouds for more than 35 years. The problem is that the "tracer" presumably influences the cloud microphysics and, in turn, the cloud dynamics. This feedback leads to a paradox. If a suitable tracer is inert to the in-cloud physical and chemical processes, then it is not a reliable tracer of the very processes needing quantification. On the other hand, if the tracer enters fully into the thermodynamic system that is the cloud, the result may be an inadvertent modification of the cloud itself.

VII. SUMMARY

The major points brought out are enumerated here.

A. Water Vapor Availability
  - Quantify the source apportionment between evaporation from water surfaces and transpiration from land surfaces.

B. Water and Ice Nuclei
  - Quantify source differences between natural and anthropogenic.
  - Ascertain the chemical evolution of nuclei from their source to their ultimate deposition.
  - Quantify the nucleating efficiency of aging aerosol.

C. Atmospheric Motions
  - Determine the filter efficiency of cloudy air for both particles and gases.

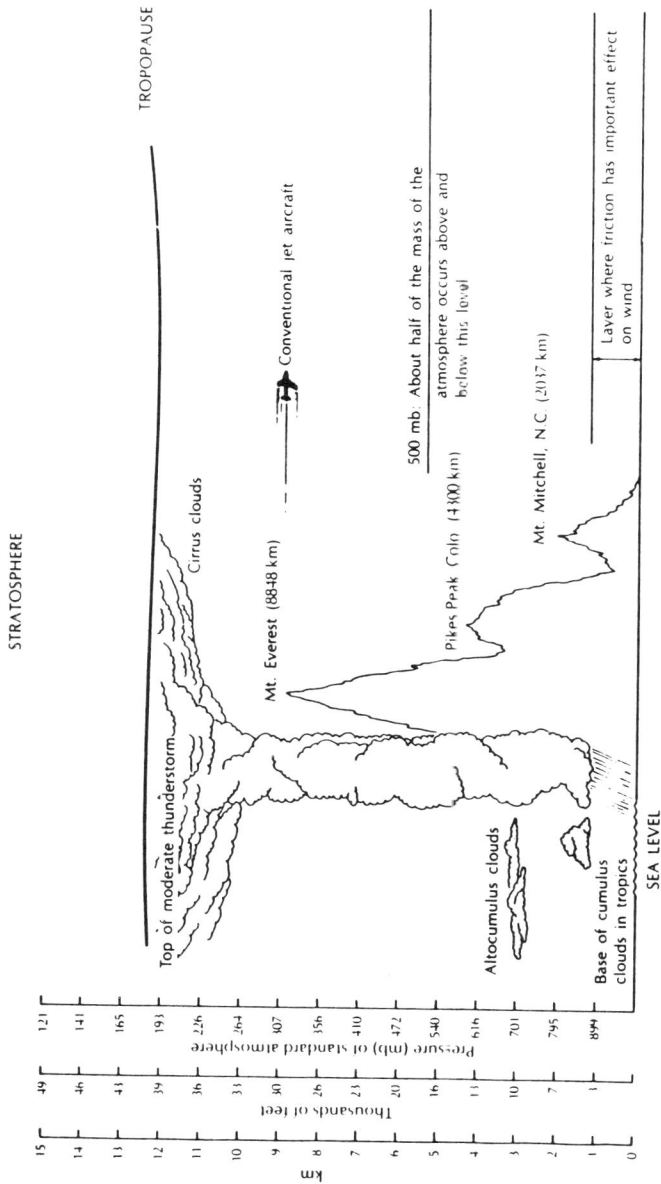

Fig. 8.   A relative vertical scale of clouds, prominent land features, and important strata of the atmosphere (Ref. 11). The similarity between the moderate thunderstorm extending into the boundary layer and a rising plume from an emission source is important to consider for understanding the distribution of gases and aerosol in the troposphere (and, perhaps, the stratosphere).

There are numerous additional problems that could be conceptualized here; for example, the influence of cloud electrification on nucleating efficiency, cloud dynamics, and cloud chemistry. However, these few general points are enough to set the framework for in-depth development of the subject.

Many of these points are not new to the atmospheric sciences community. There are some additional fruitful areas of research.

- "It is therefore essential that flight measurements be made to determine which precipitation process operates under various conditions and to obtain a quantitative verification of the operation of the assumed processes." (Ref. 13)

- "The growth of nuclei, especially at humidities of less than 70 percent, and the associated problem concerning the physical structure of nuclei (mixed nuclei, supersaturation of solution, and crystallization)." (Ref. 5)

- "The chemical composition of the aerosol. Parallel measurements of the tracers of substances and of the number of nuclei, dust particles, and haze droplets should be made. The proportion of traces of gaseous materials should be determined." (Ref. 5)

This list could go on and on, but the point is that the early post-World War II understanding of precipitation processes has not advanced significantly in spite of the technological advances through the years. One of the major impediments to a new level of learning is the necessary instrumentation for critical measurements on a time and space scale appropriate to the precipitation process. As examples, radar is extensively used to estimate internal cloud motions, but the measurements in three dimensions require periods of time that preclude simultaneity of observations of whole-cloud dynamics. In contrast to this remote sensing, aircraft measurements of particle size distribution only sample a microscopic volume of the cloud under scrutiny, and to generalize such measurements to the whole cloud can lead to serious misinformation about the chemical-thermodynamic forces at work in the system.

Without question, tracers are much needed to inquire quantitatively into the physical precipitation process and its importance to atmospheric chemistry.

REFERENCES

1.   Huff, F. A., and G. E. Stout, 1951: A preliminary study of atmospheric-
      moisture - Precipitation relationship over Illinois.
      Bull. Am. Meteor. Soc., 32, 295-297.

2.   Byers, H. R., 1965: Elements of Cloud Physics. University of Chicago
      Press, Chicago, 39-66.

3.   Semonin, R. G., 1960: Artificial precipitation potential during dry
      periods in Illinois. Physics of Precipitation, Geophys. Monograph No. 5,
      Am. Geophys. Union, Waverly Press, Baltimore, 424-431.

4.   Mason, B. J., 1957: The Physics of Clouds. Oxford University Press,
      London, 481 pp.

5.   Junge, C. E., 1951: Nuclei of atmospheric condensation.
      Compendium of Meteorology, American Meteorological Society, Boston,
      182-191.

6.   Junge, C. E., 1963: Air Chemistry and Radioactivity. Academic Press,
      New York, 328 pp.

7.   Fletcher, N. H., 1952: The Physics of Rainclouds. Cambridge
      University Press, London, 386 pp.

8.   Newton, C. W., 1969: Morphology of thunderstorms and hailstorms as
      affected by vertical wind shear. Physics of Precipitation,
      Geophys. Monograph No. 5, Am. Geophys. Union, Waverly Press,
      Baltimore, 339-347.

9.   Fujita, T. 1955: Results of detailed synoptic studies of squall lines.
      Tellus, 7, 405-436.

10.   Ochs, H. T., and R. G. Semonin, 1979: Sensitivity of a cloud
      microphysical model to an urban environment. J. Appl. Meteor., 18,
      1118-1129.

11.   Anthes, R. A., J. A. Panafsky, J. J. Cahir, and A. Rango, 1975:
      The Atmosphere. Merrill Publishing Co., Columbus, 339 pp.

12.   McDonald, J. E., 1958: The physics of cloud modification.
      Advances in Geophysics, 5, 223.

13.   Houghton, H. G., 1951: On the physics of clouds and precipitation.
      Compendium of Meteorology, American Meteorological Society, Boston,
      165-181.

SEASONAL AND DIURNAL EFFECTS ON POLLUTION TRANSPORT IN THE ROCKY MOUNTAIN WEST

Elmar R. Reiter

Department of Atmospheric Science
Colorado State University
Fort Collins, CO 80523

ABSTRACT

The effects of seasonal and diurnal variability of the heat budget of the western United States on pressure and flow systems in the planetary boundary layer (PBL) are described using data for the period 1970 to 1980 for seasonal variability and for the period 1975 to 1980 and 1982 for diurnal variability. Even in these climatic means, several mesoscale features appear in the PBL structure that demonstrate the important impact of mountains and plateaus on the flow fields and the precipitation distribution. Local-scale circulation patterns reflect the channeling and heat-source distributions associated with individual mountain ranges. These effects will have to be considered in all aspects of pollution transport and wet removal processes.

I.   INTRODUCTION

Numerous models have been developed that allow us to trace the transport of pollutants in either a Lagrangian or an Eulerian coordinate system. These models have been applied mostly over the eastern United States and Europe. Terrain effects have been included in some modeling approaches[1] but are usually ignored when long-range transport processes with a duration of several days are considered. Diurnal heating and cooling effects contribute to land- and sea-breeze reversals and to slope winds in the Pielke-Mahrer[1] model.

Because of the relatively low mean elevation of the terrain over the eastern United States, the solenoidal fields produced by the diurnal cycle of heating and cooling are not very strong. Typically, they lead to the formation of mountain- and valley-breeze systems that operate on local scales. Sea- and land-breeze effects also are of a relatively local character.

Both mountain- and seashore-related effects can have an important impact on pollution transport, especially when the large-scale gradient winds are weak and these local wind systems induce a "pendulum" motion of polluted air masses.

Over the mountainous terrain of the western United States, the solenoidal fields produced by the heating and cooling of the plateau regions are, on the average, much stronger because the difference in elevation between the plateau and the adjacent plains to the east and the Pacific Ocean to the west is much larger. It has been demonstrated by aircraft measurements during ALPEX[2] that surface soil temperatures are essentially independent of altitude but are a function, mainly, of atmospheric transmissivity, soil condition, albedo and the angle of incoming solar radiation. Thus, a horizontal surface of a plateau at high elevation will heat approximately as much as a similar surface in the low-lying plains. The horizontal difference between surface temperatures of a plateau (which is not covered by ice or snow) and the free air at the same altitude over the plains, therefore, tends to increase with increasing plateau height. This difference is especially large during the day and tends to be reversed during the night.

Furthermore, because the average temperature of the planetary boundary layer (PBL) tends to be lower over the plateau during daytime than over the plains, the vertical heat flux between the ground and the atmosphere also tends to be more effective over the plateau than over the plains.

These plateau effects lead to certain peculiarities in atmospheric structure and wind systems that are of great importance for understanding pollution transport processes over elevated, mountainous terrain.

II.   SEASONAL EFFECTS

From studies carried out over Tibet,[3] one finds that the atmosphere over the plateau tends to have the characteristics of a heat sink from October to March and those of a heat source from April to September. Part of that heat source comes from the release of latent heat which, over eastern Tibet, contributes as much to the atmospheric heat source as does sensible heat. Over the dry regions of western Tibet, the latent-heat contribution is less than 20% of the sensible heating during summer.

Over the western United States, we are lacking systematic, quantitative estimates of the magnitude of sensible and latent heat sources as a function of season. A recent study by Tang and Reiter[4] revealed that the 850-mb temperatures, characteristic of the PBL over the western United States, are warmer in summer and colder in winter than those over the adjacent ocean and plains (Fig. 1) which suggests that this plateau plays a role of seasonal heat and cold sources similar to that of the Plateau of Tibet. The result of this seasonal heat source distribution is a monsoonal change in the pressure systems that develop in the PBL. Anticyclonic systems with outflow predominate in winter, as do cyclonic systems with inflow during summer (Fig. 2).

These pressure systems, with their seasonal variability, leave a strong imprint on flow patterns and precipitation systems. For instance, the anticyclonic shear line, which is evident over Wyoming (Fig. 2b), tends to delimit the northern extent of the development of strong convective precipitation systems caused by the North American summer monsoon. Convective systems over Montana and Idaho tend to be associated with frontal disturbances enhanced by orography.

The extent of the monsoon flow can be measured as the height, or thickness, of the layer in which the monthly resultant winds undergo a direction change between winter and summer of at least $120°$ (Fig. 3). This "monsoon region" tends to surround the U.S. Plateau on three sides in the form of a crescent. A similar, crescent-shaped configuration is obtained if one looks at the layer in which the wind direction has a predominantly southerly component during summer (Fig. 4). Two regions of moisture transport become apparent from such a presentation: one over Texas and Oklahoma, which depicts the effects of the well-known low-level jet stream, and the other over Nevada, which indicates a moisture source in the Gulf of California.

The seasonal precipitation distribution over the western United States is strongly affected by the North American monsoon.

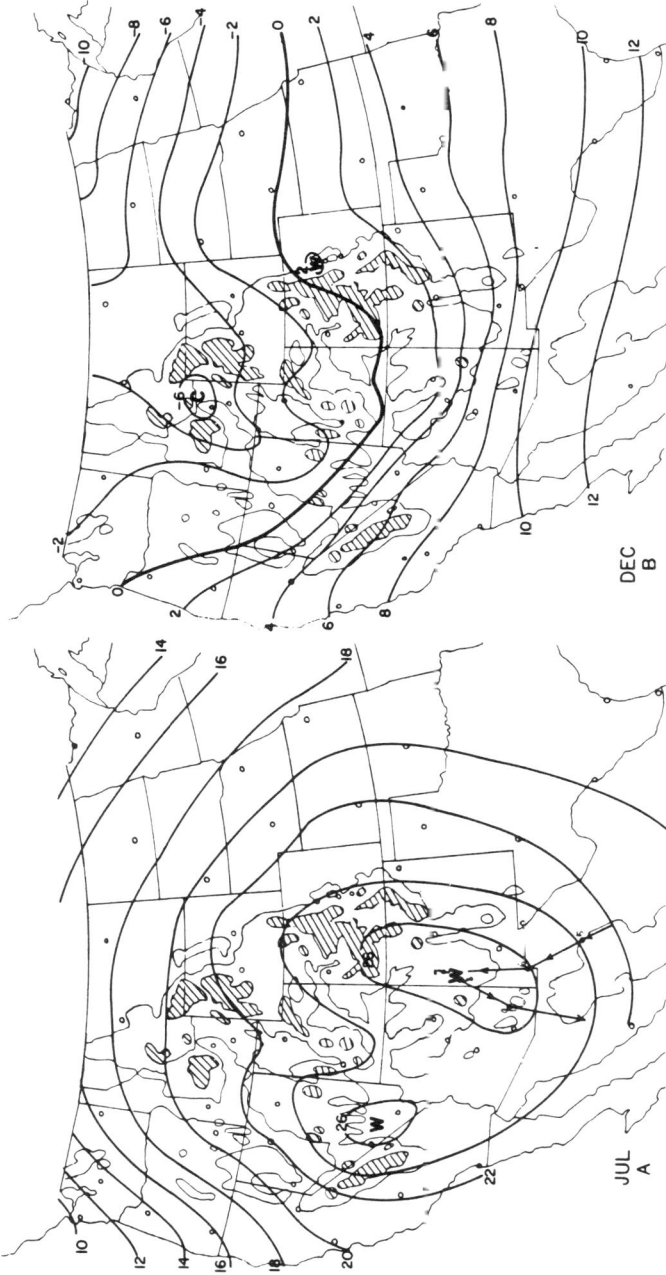

Fig. 1. Monthly mean temperature in °C at 850 mb for July and December. The 2000-m terrain contour is indicated by a thin line; terrain above 2750 m is marked by hatching. The thin line with arrows indicates the month-to-month migration of the major warm center.[4]

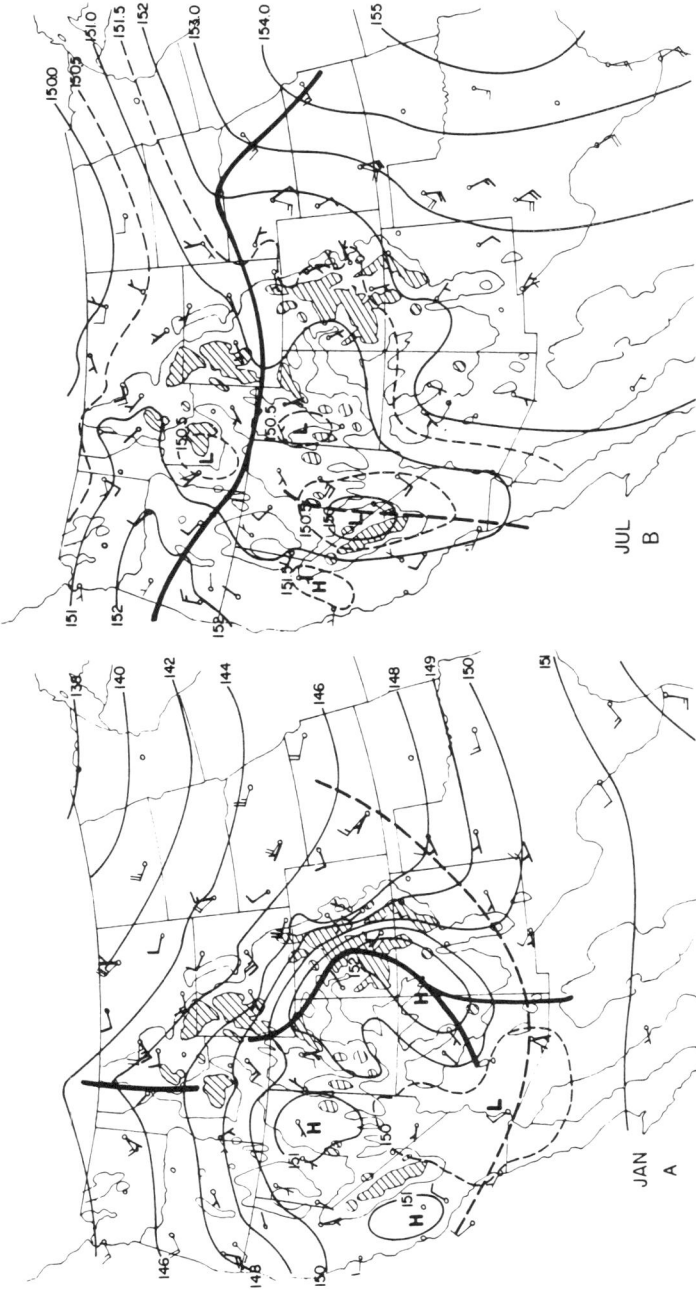

Fig. 2. Mean 850-mb heights (thin, solid lines, geopotential decameters) and resultant winds at the surface (if only one arrow is drawn); at the surface and 850 mb (if two arrows are drawn, the lighter one refers to 850 mb). The velocities are indicated as follows: No barb 0.5 m/s; short barb 0.5 to 1.4 m/a; long barb 1.5 to 2.4 m/s, etc. Heavy, full lines indicate axes of high pressure, dashed lines are axes of low pressure systems.[4]

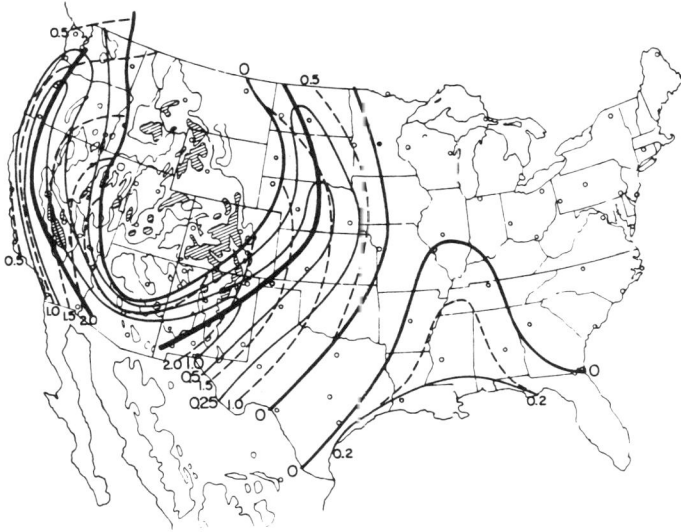

Fig. 3.   Dashed lines: height above sea level (km) of the top of the layer with
monsoonal wind reversal 120°; full lines: thickness of that layer
(km) above terrain.   Data for 1200 GMT were used in this analyses.[4]

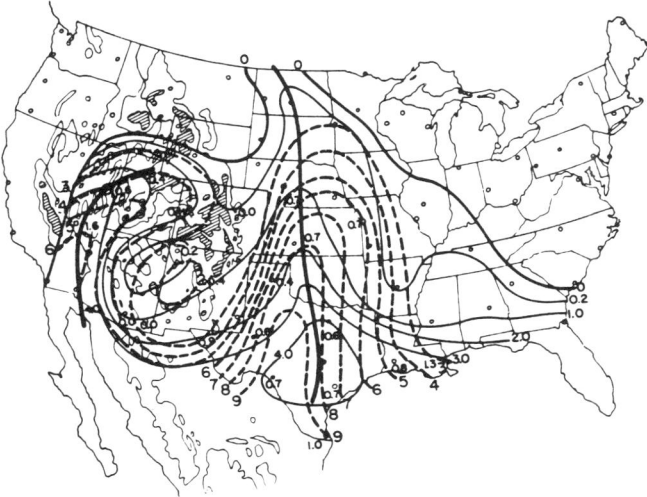

Fig. 4. Solid: Thickness of the layer (km) in which during July the resultant
         wind has a direction between 135 and 225° ("south wind"); dashed:
         maximum mean speed (m/s) of "south wind" in the lowest 3 km above
         mean sea level. The number plotted next to stations indicate the
         height above ground (km) at which the wind maximum is encountered.
         Analysis pertains to 1200 GMT data.

The two aforementioned moisture sources contribute significantly to the summer precipitation maxima shown in Fig. 5b. The winter precipitation in the Rocky Mountain region is mostly affected by atmospheric disturbances arriving from the Pacific (Fig. 5a).

The monsoon effects can lead to drastically different, sea-sonal precipitation trends over a relatively short distance. Figure 6 gives, for example, the annual precipitation curve for Buena Vista ("east"), which is under the influence of the summer monsoon and the Gulf of Mexico moisture source, and for Crested Butte ("west") on the other side of the Continental Divide, which shows a winter precipitation maximum.

The monsoonal flow and precipitation patterns are of great importance in the transport and removal of pollutants. The southerly flow channel over Arizona, Nevada, and New Mexico, provided by the summer monsoon, offers a transport mechanism for SO_2 and SO_4, which can be carried from the copper smelting operations in Arizona and New and Old Mexico. Bresch et al.[5] have shown that this transport path can contribute significantly to episodes of poor visibility in Grand Canyon National Park.

Much of the summer monsoon precipitation that falls over the Continental Divide and over the mountain ranges of the Great Basin entrains air masses that are, or can be, exposed to pollution sources from mineral smelting and/or energy development operations in the southwestern United States and in Mexico. Because this type of seasonal precipitation falls mainly over the forested mountains or over the alpine tundra above the tree line, acid precipitation effects should be studied carefully for their potentially detrimental impact on a sensitive soil and vegetation environment. It is well known that acid soils with only little buffering capability are encountered in the western United States mainly in the forested mountain regions with abundant monsoonal or frontal precipitation. Unfortunately, the summer and winter precipitation does not fall over the arid valleys of the western United States where alkaline soils would regard the addition of sulfate as a "blessing from heaven."

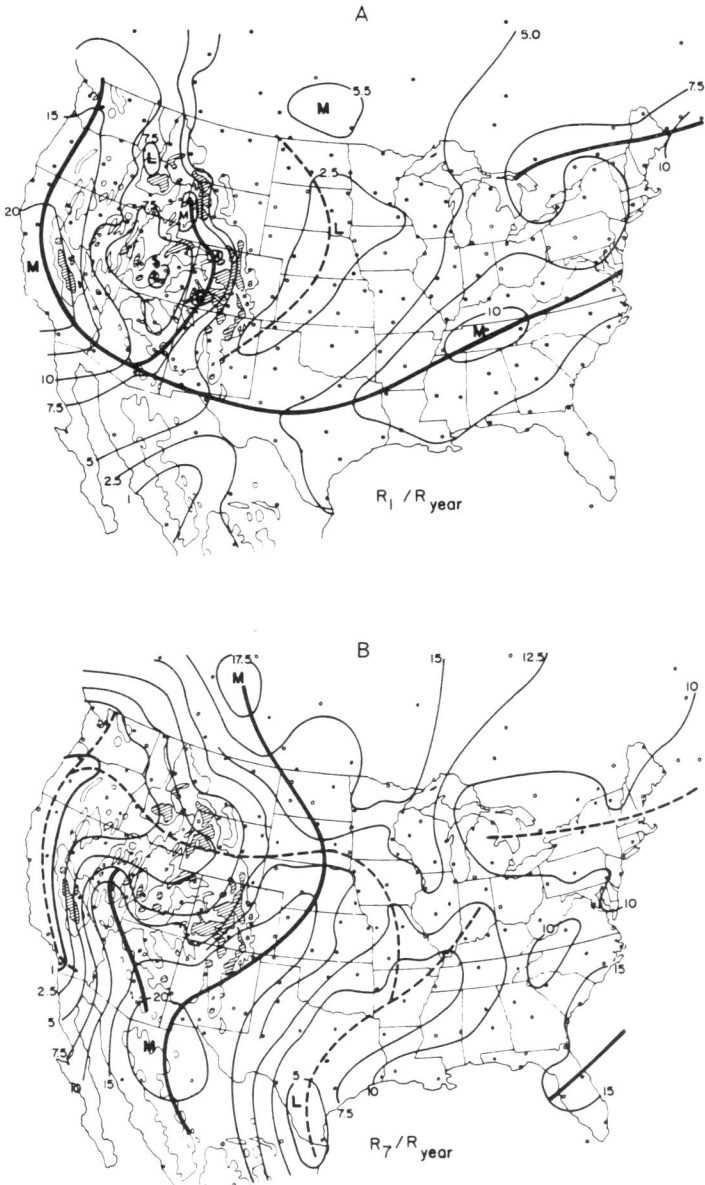

Fig. 5. Precipitation in terms of per cent of annual value for (a) January and
(b) July, based on data from 1941 to 1970. Heavy solid lines and "M"
indicate moist axes and centers; dashed lines and "L" emphasize low
precipitation regions.

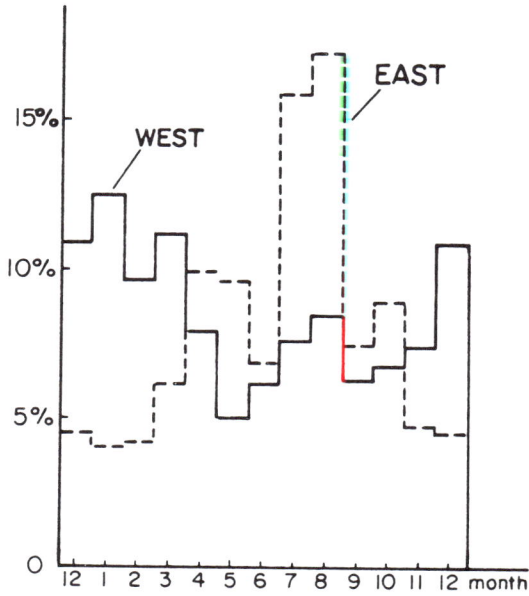

Fig. 6. Monthly precipitation, in per cent of annual, at Crested Butte (west) and Buena Vista (east), Coloradc.

III. DIURNAL EFFECTS

Reiter and Tang[6] have shown that the diurnal heating and cooling cycle over the plateau of the western United States leads to the formation of extensive heat lows during the daytime and anticyclonic systems at night (Fig. 7). Whereas the effects of these diurnal pressure changes in the PBL over the plateau on the flow systems over the Great Plains do not become very obvious from the resultant winds shown in Fig. 7, they are clearly revealed in the isallohypsic patterns and the 3-hourly changes in the resultant winds. Figure 8 shows that the low-level jet stream over Texas and Oklahoma responds significantly to the diurnally changing heat balance over the plateau to the west. From this response one can delimit the extent of a large "plateau circulation system" that varies diurnally and has a longitudinal dimension of more than 1000 km. Hence, this system will be affected by the Coriolis parameter. It is interesting to note that the U.S. Plateau and its ensuing plateau circulation are located near 30°N, where the diurnal heating cycle and the inertial oscillation have identical 24-hour periods.

The dominant effect of the plateau circulation on the diurnal evolution of convective precipitation systems becomes evident from Fig. 9. As convergent flow conditions develop during the daytime in the PBL over the plateau, the thunderstorm frequency reaches a maximum during midafternoon. With the reversal of the plateau circulation system at night, the thunderstorm activity over the plateau decreases but reaches a maximum after midnight along the convergence formed between the (reversed) plateau circulation system and the atmosphere over the Great Plains, which remains essentially unaffected by the plateau.

Again, the effects of the plateau circulation system on pollution transport and removal have to be considered as an im- portant factor over the western United States. As an example, the inflow of polluted air from the interior valley of California and from the Los Angeles region into the plateau is demonstrated in Fig. 8b, as is the entrainment of air from the eastern plains into the convective precipitation systems over the Continental Divide in Colorado and New Mexico. As has been symbolized in Fig. 7a, valleys whose orientation is "in phase" with the plateau circulation tend to develop a strong mountain- and valley-breeze system, whereas valleys with

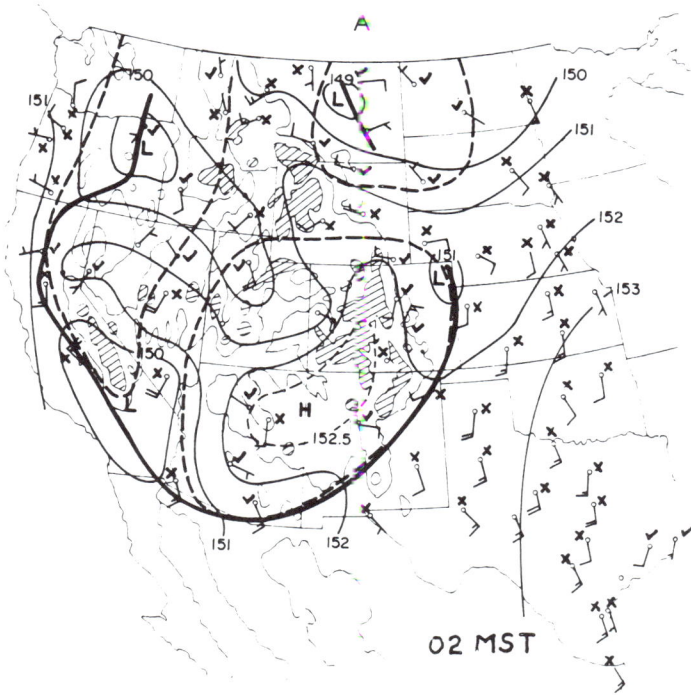

Fig. 7. Mean 850-mb heights (geopotential decameters) in July at (a) 0200
MST (0900 GMT) and (b) 1400 MST (2100 GMT).   Resultant winds are
symbolized as in Fig. 2. Trough is indicated by heavy solid line.
Stations marked "x" show no diurnal wind reversal; stations with "v"
do. Dashed line delimits the extent of the regions with diurnal
valley-mountain breeze systems. Terrain above 2750 m is hatched.[6]

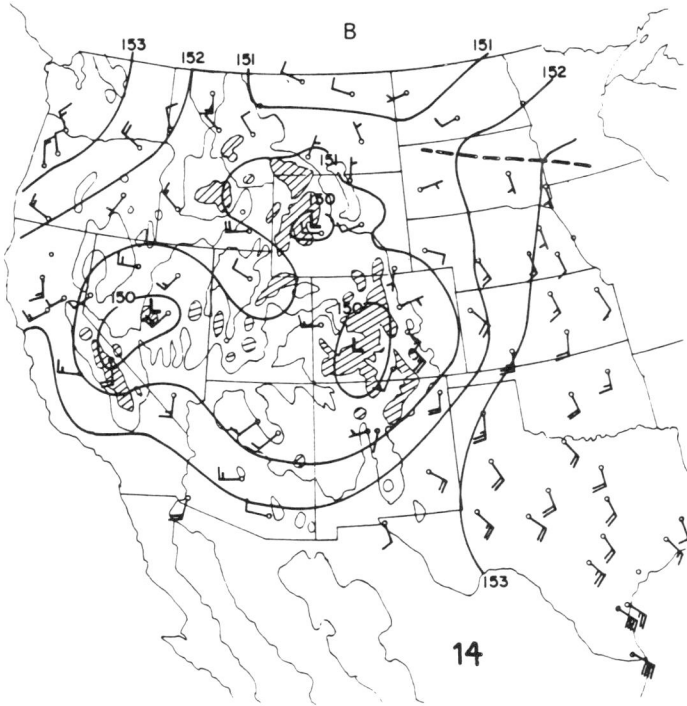

Fig. 7. (Continued)

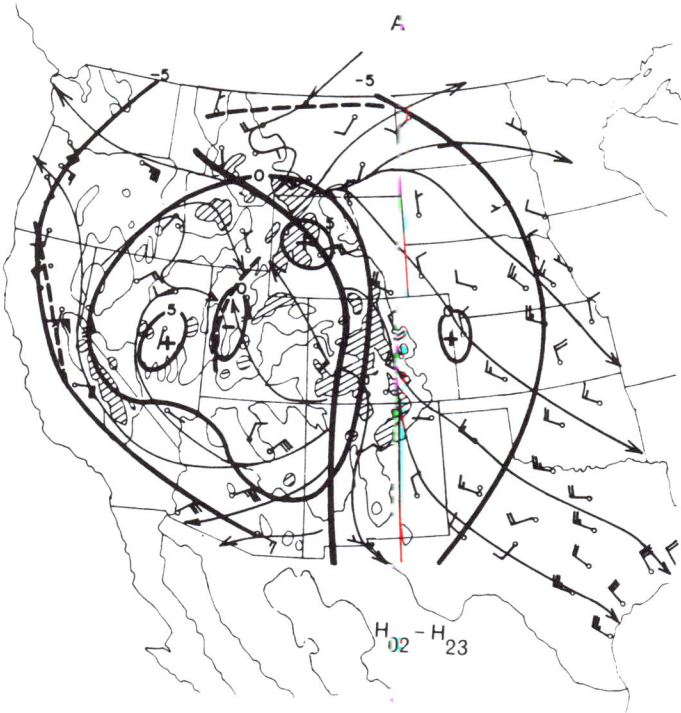

Fig. 8. Three-hour changes (a) 2300 to 0300 MST, (b) 1100 to 1400 MST, of
the 850-mb surface (in geopotential meters, solid lines of medium
thickness), 3-hour vector changes of the resultant winds (barbs on
arrows are similar to those explained in the legend to Fig. 2, but are
exaggerated by a factor of 5, i.e. a short barb stands for 0.2 m/s,
etc.), and "streamlines" of these vector changes (thin solid lines).
Axes of cyclogenetic, or convergent, and anticyclogenetic, or
divergent, centers are indicated by heavy dashed and solid lines,
respectively.[6]

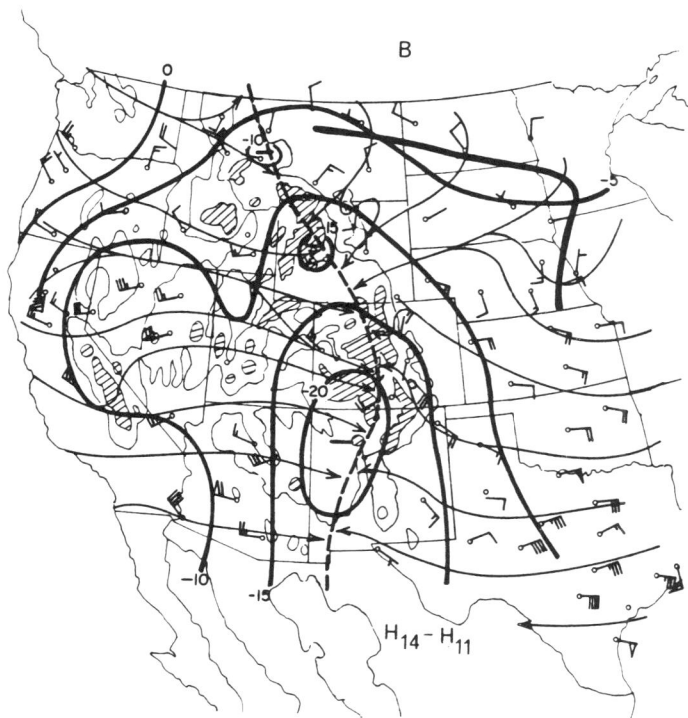

Fig. 8. (Continued)

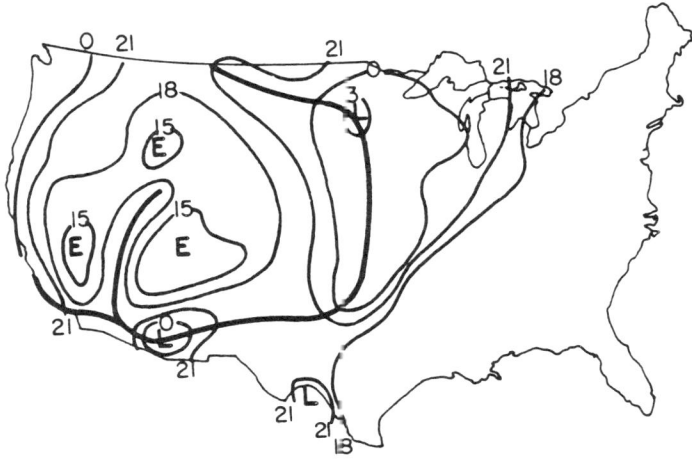

Fig. 9. Local time of the day for maximum occurrence of thunderstorm
frequency. E = early, L = late.[6] Data from Wallace.[19]

"out-of-phase" orientation show only a weak local circulation system, or none at all. These factors should be taken into account when assessing the potential impact of pollution sources in the Rocky Mountain West.

IV.   THE PLANETARY BOUNDARY AND MIXED LAYERS

Over the plains of the central United States the terms "PBL" and "mixed layer" (ML) are often used synonymously. During undisturbed weather conditions one can assume that heating of the ground during daytime produces more or less randomly distributed convection which mixes heat and momentum throughout these layers, leading to a quasi-adiabatic ML and to vertical wind profiles which, under ideal conditions, resemble those called for by Ekmans theory.[7]

Over the plateau and mountains of the western United States, random mixing no longer is an adequate assumption. Instead, slope-wind systems channel the heat received from the ground into updrafts over mountain ranges. These flow processes may be aided by synoptic pressure gradients and are also strongly affected by the "integrated" effects of the aforementioned plateau circulation system. If one takes maximum surface potential temperatures over the plateau, and the intercept of their adiabat with the local (morning) radiosonde ascent, one usually arrives at an estimate of the top of the ML of 5000 to 5500 m. Most likely, this altitude is far in excess of what one might consider to be the PBL. Reiter and Tang[6] estimated the thickness of the inflow layer (flow directed towards the plateau during the day) to be approximately 2000 m and that of the nocturnal outflow layer to be about 1500 m. These values are more consistent with what one would normally ascribe to the PBL, more so than the great thicknesses found for the ML. There are, however, considerable difficulties with both estimates.

If one of the criteria by which the PBL is characterized is the vertical transport of momentum by friction-induced mixing, the atmosphere over mountains adds a further complication. Not only are the frictional effects enhanced (which would make 2000 m rather than 1000 m an attractive value), but under the frequently occurring conditions of atmospheric structure and vertical wind profiles one can expect considerable vertical momentum transport by gravity waves. The slope-wind systems mentioned before also affect the vertical transport of momentum in a significant manner. What, therefore, is the vertical dimension of the PBL over mountains? The simple answer is: We don't know. We don't even know how to define this layer properly without violating one or the other concept that serves for its definition over the plains.

Furthermore, we have to be doubtful if the ML arrived at by Holzworth's method[8] of estimation from surface maximum temperatures and soundings can truly be considered a "mixed layer." Slope-wind assisted convection to an altitude of 5500 m most likely will encounter the condensation level and will blossom into convective cloud development over the mountain ranges. The cloud development will cut off the insolation, but the release of latent heat usually suffices to keep the convective systems going until the general reversal of the plateau circulation system dictates a widespread suppression of upward motion. Thus, the ML estimated from soundings, in reality, will reach to tropopause level in cumulus convection over the mountains and therefore cannot be considered an equivalent of the ML estimated for conditions prevailing with undisturbed weather over the plains.

As the convective systems develop over the mountains, there are compensating downward motions over the broad valleys of the Great Basin that lead to stabilization of the atmosphere and frequently to formation of subsidence inversions. What, then, is the mixed layer height over the valleys? Their high, daytime potential temperatures might not destroy any inversions but may be carried off to benefit a nearby slope circulation (Fig. 10). If one allowed for the existence of local subsidence inversions, one would be tempted to assume relatively small mixing heights over the valleys. They would not be representative, however, for the assessment of an air pollution potential because pollutants from the valley would not stagnate underneath such inversions, but would be entrained into the adjacent slope-wind circulation and subjected to washout and rainout effects.

That such local circulation systems do exist is demonstrated clearly by the vegetation distribution in the western United States. Major ridges carry significantly more vegetation, even forests, whereas the broad valleys often bear the characteristics of a desert. It has been mentioned before that the

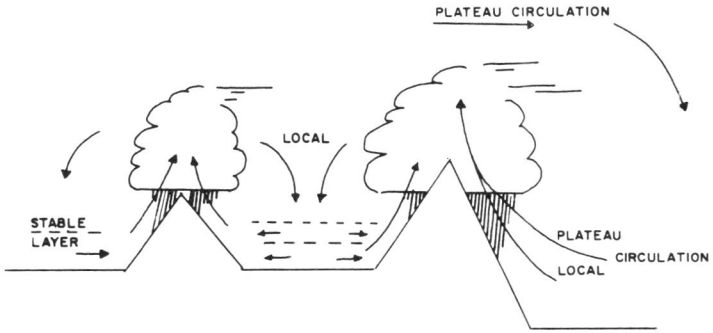

Fig. 10. Schematic view of slope-wind systems.

forest soils, and also the tundra above the tree line, are only weakly buffered against acid precipitation, whereas the valleys could easily stand the impact of acid rain but are not receiving it.

V.   LOCAL CIRCULATION SYSTEMS

The slope-wind systems described above are driven by the surface energy budget and later in the course of their development, perhaps, by the release of latent heat. Their effectiveness is demonstrated in Fig. 11, which shows the resultant wind speeds and directions as a function of the time of day in July in Denver and Colorado Springs. The two stations are separated by the rather low Palmer Ridge, which extends eastward from the Continental Divide. The diurnal wind system in Denver follows the southwest-northeast orientation of the valley of the South Platte River. Both stations reveal the upslope motions prescribed by the plateau circulation system. However, the Palmer Ridge between the two stations induces a marked convergence of flow during the day, and a divergence during the night. Local forecasters know that thunderstorms that develop over the Continental Divide during the day have a tendency to migrate eastward during the afternoon, following the convergence line induced by the Palmer Ridge. That ridge itself bears the evidence of the convective precipitation enhancement; it carries the "Black Forest," which extends eastward into the more or less barren plains.

VI.   TRAJECTORY PATTERNS

A number of computer programs are available that will construct trajectories of air motion forward or backward in time. (For a summary see Bresch et al.[5]) Our own model[5] has been adapted from a multilayer model developed by Draxler and Taylor[9] (see also Draxler[10]) and has been tested extensively by running it in a backwards mode over the southwestern United States. The model considers the four layers shown in Fig. 12. Trajectories originating at night assume that pollution produced near ground level is constrained within the lowest 300 m above terrain. During the day, mixing is allowed to proceed either to 5000 m above terrain or to the top of the mixed layer determined from soundings--whichever value is lower. Wind shear

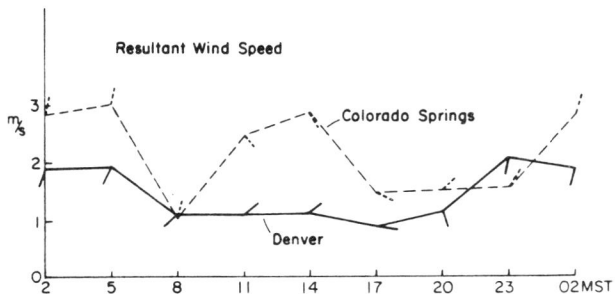

Fig. 11. Mean diurnal variation of resultant wind speed (m/s) and direction at Colorado Springs and Denver, Colorado, for July.[6]

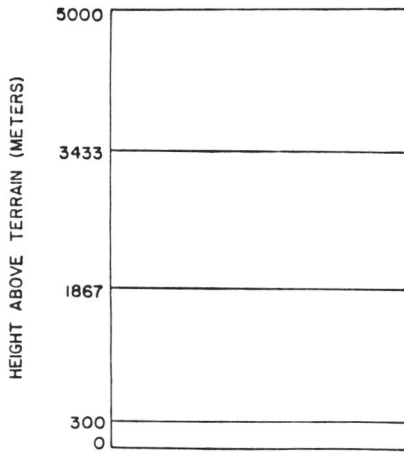

Fig. 12. Layers used in the CSU (LORMAT) trajectory model.[5]

effects are introduced by considering separate trajectories in each layer. During the night no mixing between layers is allowed to occur.

Layer-mean winds are obtained from soundings and interpolated objectively[11,12] to grid points at 3-hour intervals. The data base used so far was developed by Heffter[13] and gives winds at aerological stations in meters above terrain. Thus, terrain height is implicit in the data base, but not with all the details of topography between stations. Because radiosonde stations usually are located in valleys, the effects of mountain ranges are virtually ignored by the model.

There are other models available that take into account more of the terrain details,[14,15,16] but they are expensive to run, deal with a relatively limited domain, or do not require mass balance. Clearly, the effects of topography, and especially of the large-, meso-, and small-scale motion systems discussed above, require considerably more attention in future trajectory work.

Even without the detailed resolution of terrain, the C.S.U. model yields interesting results. Figure 13 shows the probability (in per cent times 10) that within 72-hour time intervals, backward trajectories, whose destination is the Grand Canyon measurement site, originated in a certain 1° by 1° grid area. Between September 1980 and August 1981, most of the trajectories came either from the vicinity of the Grand Canyon or from southern California in southwesterly flow.

If one considers the 72-hour trajectory origins for air with more than 375 ng/m^3 of fine (smaller than 2.5 µm in diameter) particulate sulfur measured at the Grand Canyon and divides that number in each grid square by the number of all trajectories (for clean as well as dirty air) in that square, one arrives at the conditional probability of trajectory origins shown in Fig. 14. This diagram points to possible sulfur origins in the smelting operations of northern Mexico, southern Arizona, and New Mexico and in the pollution sources of the interior valley of California.

This kind of representation does not yet take into account that if trajectories arrived at the target site with equal probability from each direction, the probability of a given source region affecting the target region would decrease proportional to the square of the distance between source

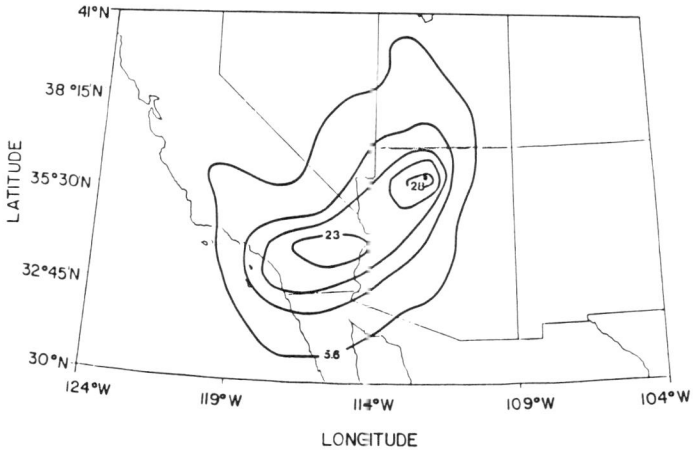

Fig. 13. Probability that a trajectory ending at Grand Canyon originated in a
10° by 10° grid box within a 72-hour period. Data from September
1980 to August, 1981.[5]

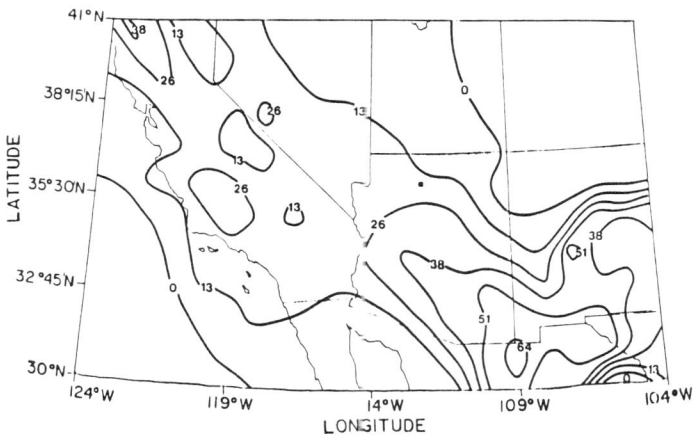

Fig. 14. Conditional probability for trajectories with sulfur 375 ng/m³ ending
at Grand Canyon, having originated in a 1° by 1° grid box.[5]

region and target region.[17] If one divides the probability of a polluted trajectory originating in a given grid square by this distance-weighted probability-density function, one arrives at the distribution shown in Fig. 15. During the year for which data were available, high-sulfur episodes at the Grand Canyon were frequently associated with trajectories originating up to 72 hours earlier in the Los Angeles area, the Phoenix region, and the interior valley of California. Because of the prevailing wind direction, the smelting operations in southeastern Arizona and in New Mexico seemed to have had little effect.

VII. CONCLUSIONS AND OUTLOOK

In the foregoing discussion, the existence of organized, nonrandom motion systems tied to the topographic features of the western United States was demonstrated. Topographic effects became apparent on seasonal and diurnal time scales and on space scales ranging from several thousand to a few kilometers. These effects will exercise a dominant control on the transport of pollutants, but also on the intensity and location of wet removal processes. A number of field experiments have been carried out to trace the transport of pollutants in complex terrain. Notably, the ASCOT Program has provided much valuable data. Other, even more ambitious tracer experiments are presently under consideration. These exercises, however, will not diminish the demand for improved trajectory techniques that are capable of digesting the meteorological and climatological data available from standard ground-based and upper-air observation networks. These techniques will have to come to grips with the fact that the barrier and heat source effects of mountain ranges and plateaus exercise a permanent control on the meso- and local climate of airflow and precipitation at scales below the resolution of the standard networks and of most mesoscale numerical models.

In principle, the problem can be considered to be rather simple. The deflection of airflow by a given topographic feature (e.g., the mountains of the Continental Divide in Colorado) should be replicable if the same conditions of flow direction, speed, and vertical shear, of atmospheric stability and humidity distribution, and of sensible heat sources at the ground prevailed in individual cases. The same cloud patterns, and the same latent

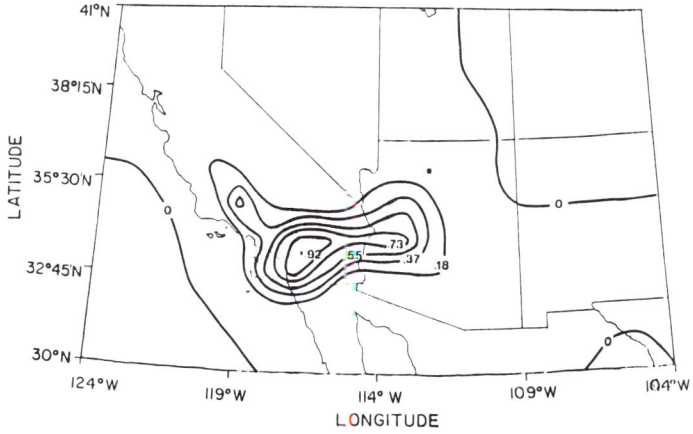

Fig. 15. "Source contribution," i.e. probability of "polluted" 72-hour trajectory origins divided by distance-weighted probability-density function.[5]

heat sources should prevail in each of these "replicable" cases. Unfortunately, nature with its infinite variety of parameter combinations will not cooperate by providing truly replicable cases. One could engage statistics to find empirical relationships between the parameters mentioned above and the resulting vertical and horizontal flow fields by using a large number of different cases as data base. One would think that such an approach should be rather straightforward and manageable. The problem is that hardly any of the data required to carry out such a study exist.

The flow conditions around mountain obstacles usually are not known in the desired detail, but they probably could be "fudged" with a certain degree of authority by employing numerical models. The sensible heat source distribution at ground level is a function of terrain orientation, soil conditions, vegetation, etc. and, in general, is not adequately known. The precipitation and latent heat source distribution around a mountain range is difficult to determine when the rain gauges are located in the valleys. This multitude of unsolved data problems might provide enough discouragement to forego any attempt at filling in the mesoscale details of pollution transport and wet removal beyond what has already been estimated grossly in the past. Even if small- and mesoscale models were made available that would take into account detailed topographic features, what is there to verify their output?

Perhaps a simpler approach can be chosen to provide some of the needed answers. In essence, we are only interested in two major aspects of topographic effects: (1) the channeling effects of mountain barriers on flow that is not well mixed over deep layers, and (2) the convergence and divergence fields generated by convection over the mountain ranges and usually resulting in mesoscale, convective precipitation systems.

The first of these two aspects can be treated in a rather simple way by employing a multilayer flow model that contains a reasonably smoothed topography, obeys the continuity equation, and disallows flow through the ground surface. Such models are currently available (e.g., Fossberg[14] and others). The second aspect is more difficult to handle but could be attacked by using a reasonably good cloud model (see e.g. Cotton and Tripoli[18]) that provides an estimate of updraft velocities under given, large-scale atmospheric conditions and with reasonable guesses on precipitation amounts.

Such velocity estimates, then, would provide estimates of the mesoscale convergence and divergence patterns caused by topography and precipitation systems. These patterns could then be superimposed upon the large-scale convergence patterns to yield detailed, hypothetical wind fields that would, at least, be consistent with some of the observed consequences, namely the orographically generated convective systems.

Although this approach might sound reasonable, it is by no means simple. We are presently working on this concept of parameterized flow modeling, but no results are available yet.

ACKNOWLEDGMENTS

The research reported here has been supported by the following research grants: Department of Energy Contract DE-AC02-76EV1340; National Science Foundation Grants ATM 82-11408 and ATM 83-13270, Climate Dynamics Program, Atmospheric Science Division; National Park Service NA-81RA-H-00001; and the Air Force Office of Scientific Research, Air Force Systems Command, USAF, under Grant Number AFOSR 82-0162. The United States Government is authorized to reproduce and distribute reprints for Governmental purposes notwithstanding any copyright notation thereon.

REFERENCES

1. Pielke, R. A. and Y. Mahrer: Representation of the heated planetary boundary layer in mesoscale models with coarse vertical resolution. J. Atmos., 32 (12), 2288-2308 (1975).

2. Bolle, H. J.: Application of meteorological satellite to mountain meteorology. Presented at the International Symposium on the Qinghai-Xizang (Tibet) Plateau and Mountain Meteorology, March 20-24, 1984, Beijing, People's Republic of China (1984).

3. Ye, Duzheng: Some aspects of the thermal influences of the Qinghai-Tibetan Plateau on the atmospheric circulation. Arch. Meteor. Geoph. Biokl., Ser. A, 31, 205-220 (1982).

4. Tang, Maocang and Elmar R. Reiter: Plateau monsoons of the Northern Hemisphere: A comparison between North America and Tibet. Accepted by Mon. Wea. Rev. (1984).

5. Bresch, J. F., L. L. Ashbaugh, T. Henmi, and E. R. Reiter: Comparison of a single-layer and a multilayer transport model for residence time analysis. The Fourth Joint Conference on Applications of Air Pollution Meteorology, October 29-November 2, 1984, Portland, Oregon, (1984).

6. Reiter, Elmar R. and Maocang Tang: Plateau effects on diurnal circulation patterns. Accepted by Mon. Wea. Rev. (1984).

7. Hoxit, L. R.: Planetary boundary layer winds in baroclinic conditions. J. Atmos. Sci., 31, 1003-1020 (1974).

8. Holzworth, G. C.: Mixing heights, wind speeds, and potential for urban air pollution throughout the contiguous United States. AG-101, Office of Air Programs, Environmental Pro- tection Agency. Research Triangle Park, NC 27711 (1972).

9. Draxler, R. R. and A. D. Taylor: Horizontal dispersion parameters for long-range transport modeling. J. Appl. Meteor., 21, 367-372 (1982).

10. Draxler, R. R.: User's guide for a long-range multilayer atmospheric transport and dispersion model. NOAA Tech. Memo., ERL ARL-112, 10 pp. (1982).

11. Barnes, S. L.: A technique for maximizing details in numeri- cal weather map analysis. J. Appl. Meteor., 3, 396-409 (1964).

12. Barnes, S. L.: Mesoscale objective map analysis using weighted time-series observations. NOAA Tech. Memo., ERL NSSL-62, 60 pp. (1973).

13. Heffter, J. L., 1981: Air Resources Laboratories Atmospheric transport and dispersion model (ARL-ATAD). NOAA Tech. Memo., ERL ARL-81, 21 pp]NTIS PB80-163652[.

14. Fossberg, M. A., W. E. Marlatt, and L. Krupnak: Estimating airflow patterns over complex terrain. USDA Forest Service Research Paper RM-162, Rocky Mountain Forest and Range Ex- periment Station, Fort Collins, CO, 16 pp. (1976).

15. Davis, C. G. and S. S. Bunker: Mass-consistent windfields - July 22 geyser's area. Los Alamos Scientific Laboratory Report LA-UR-80-1092, Presented at the ASCOT Gettysburg Meeting, April 15-18 (1980).

16. Mass, Clifford F. and David P. Dempsey: A one-level mesoscale model for diagnosing or short-term forecasting surface winds in complex terrain. Presented at the International Symposium on the Qinghai-Xizang (Tibet) Plateau and Mountain Meteorology, March 20-24, 1984, Beijing, People's Republic of China, (1984).

17. Ashbaugh, L. L., W. C. Malm, and W. Z. Sadeh: A methodology for establishing the probability of the origin of air masses containing high pollution concentrations. Presented at the 76th Annual APCA Meeting, Paper No. 83-10P.13, Atlanta, GA, June (1983).

18. Cotton, W. R. and G. J. Tripoli: Cumulus convection in shear- flow three-dimensional numerical experiments. J. Atmos. Sci., 35, 1503-1521 (1978).

19. Wallace, J. M., 1975: Diurnal variations in precipitation and thunderstorm frequency over the conterminous United States. Mon. Wea. Rev., 103, 406-419.

Part V

Ad Hoc Presentations

SOIL AEROSOLS AS TRACERS OF LONG-RANGE TRANSPORT

Joseph M. Prospero

University of Miami
Rosenstiel School of Marine and Atmospheric Science
Division of Marine and Atmospheric Chemistry
Miami, Florida

John Merrill

Graduate School of Oceanography
University of Rhode Island
Kingston, Rhode Island

ABSTRACT

In many ocean regions, a major aerosol constituent is
mineral matter derived from the continents. The greatest
concentrations of soil aerosol particles are found over marine
areas "downwind" from arid regions and deserts. Because of the
transport of soil material out of North Africa, the Arabian
Peninsula, Asia, and India, the mean mineral aerosol
concentrations can be quite great over the tropical North
Atlantic, the Indian Ocean, the Mediterranean, and the North
Pacific. The concentration of soil aerosols in these regions
produces highly turbid sky conditions. Dust storms are often
visible in satellite imagery, and remote sensing techniques can be
used to follow the movement of the dust clouds.

I.   INTRODUCTION

Continental areas cover only about one third of the surface of the
Earth, and about one third of the continents is classified as arid. Many of
these arid regions are major sources of soil dust, which can be transported
great distances by winds. The subject of soil aerosols has been reviewed in a
number of recent papers[1-5]. These reviews show that the concentration of
soil dust in the atmosphere is highly variable on a geographical and temporal
basis.

A wide variety of meteorological factors determine the frequency of
dust storms, the altitude to which dust is lifted, and the subsequent transport
and dispersion of the dust cloud. Because dust storm generation is usually
associated with energetic meteorological events, the generation process will
be quite complex. The subsequent transport path may also be complicated.
For example, major dust storms occur in Asia during the late winter and

spring[6-8]. High concentrations of Asian dust are measured over the North
Pacific at this time of year even at sites that are situated in a relatively
deep easterly flow. We conclude that the dust is transported in the wester-
lies, primarily at higher levels; the dust-laden air parcels eventually sink and
enter the easterlies, where they are detected at ground-level stations[9].

Large dust storms that originated over the Sahara are often observed
over the tropical North Atlantic. Meteorological conditions produce a
layered structure in these dust outbreaks. Saharan dust events seem to be
associated with deep, hot, convectively generated layers of air that have a
constant potential temperature throughout their depth, some 5 to 7 km[10].
These parcels move westward aross the Sahara and emerge over the Atlantic
where they are undercut by the relatively cool, moist trade winds, which have
a strong northerly component along the coast. Consequently, a very strong
inversion is generated at the base of the Saharan air layer. As a rule, the
concentration of dust is much higher above this inversion than below it;
conversely, sea salt aerosol is confined below the inversion. This vertical
structure is preserved as the dust outbreaks traverse the Atlantic into the
Caribbean and even to the southeast coast of the United States[10].

Data on the concentration of soil dust in the marine atmosphere are
compiled in Table I (Ref. 3). Included for comparison in the table are values
for two nonmarine cases. The impact of soil aerosols from arid regions is
clearly evident; the largest concentrations are found in the tropical North
Atlantic, the North Pacific, the Mediterranean, the Indian Ocean, and the
Bay of Bengal. However, we point out that there is very little data on
concentrations in the South Atlantic, the Indian Ocean, and the South Pacific
(i.e., most of the Southern Hemisphere).

In the following sections we present data that illustrate the temporal
and areal variability of soil aerosols in relatively remote marine sites.

II. TROPICAL NORTH ATLANTIC

The longest running record of soil aerosol measurements is that
obtained in an aerosol program that has been in effect at Barbados, West
Indies, since 1965 (Refs. 11, 3). The monthly mean mineral aerosol
concentration data for the period up to the end of the 1980 are shown in
Fig. 1.

TABLE I

MINERAL AEROSOL CONCENTRATIONS IN THE ATMOSPHERE[a](Ref. 3)

	Geometric Mean Conc.			Range		n
	C[b]	C(Al)[c]	C(Fe)[d]			
North Atlantic						
Northern Norway (coast)		0.56	1.32	0.08	2.63	21
Lerwick, Shetland Island (coast)		0.84	1.84	0.28	4.21	11
Bermuda (island)		1.96	2.37	0.04	50.00	29-60
Central and Northern (ship)	0.36			50.02	14.1	109
Eastern Tropical (ship)		0.70	1.05	0.17	2.90	8
Tropical and Equatorial (ship)	14.2			1.15	186.7	22
Tropical (Islands)	14.6			0.36	199.5	149
South Atlantic						
Gulf of Guinea (ship)			3.16	2.11	4.47	9
Tropical and Central (ship)	0.69			0.04	7.52	35
Pacific						
Oahu, Hawaii (island)			0.24	0.03	1.32	56-119
28°N-40°S (ship)	0.35			0.05	2.34	24
Indian Bay of Bengal						
7°N-15°S (ship)	4.76			0.49	11.4	5
Mediterranean (ship)	4.29			2.76	9.50	13
MS, SC, and PS (ship)	1.09			0.24	3.89	6
Urban		22.45	44.74	4.77	126.34	
South Pole		0.008	0.013	0.003	0.026	4-10

---

a   Units: 10^{-6} g/m^3 air (STP).
b   Bulk isol. aerosol concentration.
c   Concentration based on AL concentration.
d   Concentration based on Fe concentration.
e   Malacca Straits, South China, and Philippine Seas.

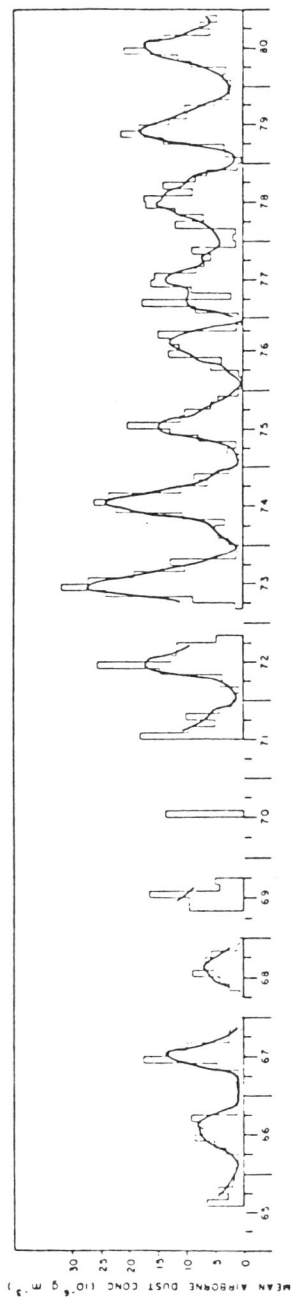

Fig. 1.   Mean monthly mineral aerosol concentration at Barbados, West Indies.

Several trends are evident in these data. First, there is a very marked seasonal variation in the dust transport, the maximum occurring in summer months. Second, there are some long term trends in the mean summer values, the most obvious being a pronounced maximum that occurred in the early 1970 at the time of the severe drought in North Africa. The mean monthly summer values decreased markedly for several years after 1973 and 1974. However, more recent evidence indicates that the concentration at Barbados (and hence the dust transport out of Africa) is once again increasing. This increase may be a consequence of the subnormal rainfalls that have been recorded in West Africa and the Sahel for the past several years.

The seasonal cycle in dust concentration at Barbados apparently is not due to any cessation of dust transport out of Africa during the winter months. Indeed, the frequency of occurrence of dust haze along the west cost of Africa is significantly greater in winter than during the summer[3]. The seasonal cycle of dust concentration of Barbados could be interpreted as being a consequence of the well-known seasonal southward shift of the large-scale circulation patterns over Africa and the tropical Atlantic. This hypothesis has been substantiated by aerosol data from a sampling station in Cayenne, French Guiana (4°50'N, 52°22'W); the station was in continuous operation from December 1977 to December 1979 (Ref. 12). The greatest mineral aerosol concentrations are observed in the early months of the year with the maximum monthly mean occurring in March in both 1978 and 1979. The Cayenne annual maximum values are markedly higher than the annual maximum monthly means for Barbados.

III. NORTH PACIFIC

Mineral aerosol studies have been made in a network of seven stations in the North Pacific as a part of the Sea/Air Exchange (SEAREX) Program. Beginning in early 1981, aerosol samples were collected continuously for 1-week periods[9]. The mineral aerosol concentration data are presented in Fig. 2.

The seasonality of dust transport is clearly evident from the data for Shemya, Midway, Oahu, and Enewetak. This same seasonal trend is also evident in a 4-year record (1979 to 1982) of soil aerosol measurements made at Mauna Loa Observatory on Hawaii[13]. This data record is especially

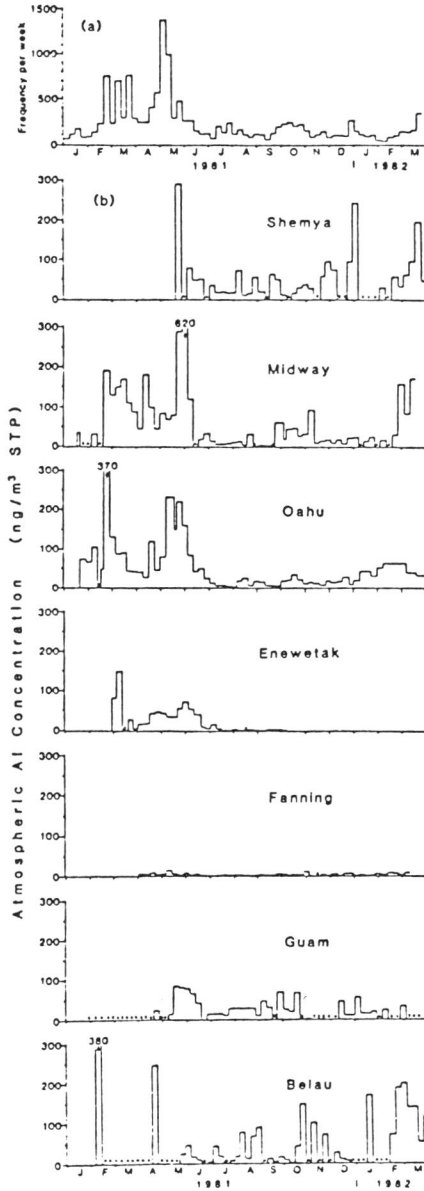

Fig. 2. (a) Frequency of dust storm reports in Asia.
(b) Temporal variation of the atmospheric Al concentration in the
SEAREX North Pacific Network. The mineral aerosol concentration
can be computed by multiplying the Al concentration by 15.

interesting because the aerosol measurements were made above the trade wind inversion.   Note that the stations at Shemya, Midway, Oahu, and Enewetak cover a latitude span from 52°N to 11°N.   In contrast, the dust concentrations at Fanning are consistently low because of its location to the south of the intertropical convergence zone (ITCZ).

The transition from high dust period to low dust period is synchronous across most of the North Pacific[9].   Also, the time of transition appears to be relatively constant from year to year.   This can be seen by comparing the Enewetak dust concentration data for May, June, and July during the 1979 SEAREX field experiment with those obtained at the Enewetak dust network site in 1981[9].   This seasonal pattern is grossly similar to the frequency of dust storm reports in Asia for 1981, data that we collect and collate in our program.   This seasonal variation is also similar to that of documented dust events ("Kosa") in Japan.   Kosa is yellow or brown airborne dust that is transported from the Asian continent.   Scenarios for a number of major dust episodes are discussed in Ref. 9.

On the basis of the measured mean dust concentrations, we have estimated the deposition rates to the North Pacific[9].   The annual atmospheric flux of crustal material to the surface of the North Pacific is comparable both to the present day downward flux of alumino-silicate minerals in the water column in the North Pacific as measured by sediment traps (50 $\mu g/cm^2/yr$) and to sediment accumulation rates of nonbiogenic components in the North Pacific (50 to 200 $\mu g/cm^2/yr$).

IV. CONCLUSIONS

Studies carried out thus far clearly show that long-range soil dust transport occurs in many ocean regions.   These episodes are synoptic scale events and they occur relatively frequently; they are not anomalous occurrences.   The dust concentrations are sufficiently great that they can be routinely detected in satellite imagery[3].   The high dust concentrations result in reduced atmospheric transmission, which can be readily measured with simple optical devices at ground level[14].   All these factors suggest that soil aerosols can serve as excellent tracers of atmospheric transport processes.

REFERENCES

1.   Goudie, A. S., 1978, Dust Storms and Their Geomorphological Implica-
     tions, J. Arid Environ. 1:291-311.

2.   Prospero, J. M., 1979, Mineral and Sea Salt Aerosol Concentrations in Various Ocean Regions, J. Geophys. Res. 84:725-731.

3.   Prospero, J. M., 1981, Eolian Transport to the World Ocean, In The Oceanic Lithosphere: The Sea, C. Emiliani (Ed.), Vol. VII, Wiley Interscience, New York, pp. 801-874.

4.   Morales, C. (Ed.), 1979, Saharan Dust: Mobilization Transport, Deposition, SCOPE 14, Wiley and Sons Ltd., Chichester, 297 pp.

5.   Pewe, T. (Ed.), 1981, Desert Dust: Origin, Characteristics, and Effect on Man, Special Paper 186, Geological Society of America, Boulder, Colorado, 303 pp.

6.   Duce, R. A., C. K. Unni, B. J. Ray, J. M. Prospero, and J. T. Merrill, 1980, Long-Range Atmospheric Transport of Soil Dust from Asia to the Tropical North Pacific: Temporal Variability, J. Geophys. Res. 88:5321-5342.

7.   Avila, L., J. T. Merrill, and R. Bleck, 1984, Asian Dust Storms and the Mechanism of Dust Injection Into the Troposphere, Monthly Weather Review (Submitted).

8.   Merrill, J. T. and R. Bleck, 1984, Modelling Atmospheric Transport to the Marshall Islands, J. Geophys. Res. (Submitted).

9.   Uematsu, M., R. A. Duce, J. M. Prospero, L. Chen, J. T. Merrill, and R. L. McDonald, 1983, The Transport of Mineral Aerosol from Asia Over the North Pacific Ocean, J. Geophys. Res. 88:5343-5352.

10. Prospero, J. M. and T. N. Carlson, 1981, Saharan Air Outbreaks Over the Tropical North Atlantic, Pure Appl. Geophys. 119:677-691.

11. Prospero, J. M. and R. T. Nees, 1977, Dust Concentration in the Atmosphere of the Equatorial North Atlantic: Possible Relationship to the Sahelian Drought, Science 196:1196-1198.

12. Prospero, J. M., R. A. Glaccum, and R. T. Nees, 1981, Atmospheric Transport of Soil Dust from Africa and South America, Nature 289:270-572.

13. Parrington, J. R., W. H. Zoller, and N. K. Aras, 1983, Asian Dust: Seasonal Transport to the Hawaiian Islands, Science 220:195-198.

14. Prospero, J. M., D. L. Savoie, T. N. Carlson, and R. T. Nees, 1978, Monitoring Saharan Aerosol Transport by Means of Atmospheric

Turbidity Measurements, <u>In</u> Saharan Dust: Mobilization, Transport, Deposition, Scientific Committee cn Problems of the Environment, Int. Council of Scientific Unions, Publication 14, John Wiley and Sons, Ltd., Chichester, pp. 171-186.

SOME REMARKS ON CONTINENTAL AND GLOBAL TRACERS

Kenneth A. Rahn

Graduate School of Oceanography
University of Rhode Island
Narragansett, Rhode Island

During the first day of this workshop, little mention was made of continental and/or global tracers of any type, in spite of the fact they have rapidly increased in quantity and quality during recent years. I am happy to have this chance to offer a few remarks on our experiences in this area.

Large-scale tracers may be natural or anthropogenic. Natural tracers tend to represent classes of sources, e.g., desert dust, rather than specific sources. To date, specific natural sources such as individual deserts have been usually identified by default in cases where no other similar sources are near enough to account for the material found.

Anthropogenic tracers can be deliberate or inadvertent. Deliberate tracers are injected into the atmosphere by the investigator and are followed as the air moves. Inadvertent tracers (also called "adventitious" or "tracers of opportunity" at this workshop) are substances that are in the atmosphere already and are exploited for tracer purposes after the fact. Continental and global tracers fall into both these categories.

Tracers may have two quite different uses: to follow the motions of air or to determine the origin of a particular substance. Historically, bomb debris served the former purpose on the continental and global scale, but it is much less useful now that atmospheric testing has largely ceased. Current needs for large-scale tracers are more oriented toward understanding the sources and transport of pollution substances, although many pure meteorological needs remain. As transport of particulate and gaseous air pollutants on continental, hemispheric, and global scales becomes more widely recognized, new tracers are needed. Bomb debris, even if it were still present, would not be suited to this purpose because it starts from the stratosphere or upper troposphere and works its way downward, whereas air pollution starts at the surface and works its way upward. Tracers are particularly important on scales larger than synoptic because that is where trajectory analysis and dispersion calculations are still unreliable. Although long-term circulation is fairly well understood on this scale, short-term motions of individual parcels of air are still highly uncertain.

Deliberate large-scale tracers are highly desirable because they can be used under controlled circumstances. However, they are difficult to use because of the large masses, costs, and numbers of samples required and because of the logistical problems associated with tracking large-scale motions. Another problem, at least with deliberate tracers such as the perfluorocarbons, is that their chemical properties are very different from the species that they trace: the very inertness that makes these tracers conservative and easy to follow decreases their resemblance to the more reactive systems (e.g., sulfur) that they are trying to trace. In other words, there is an unusual premium associated with finding inadvertent tracers for the large scale. Fortunately, inadvertent tracers have evolved rapidly in the last few years. Unfortunately, this rapid evolution has rendered them unknown and unappreciated outside the immediate circle of researchers who deal with them.

At the moment, there are two principal types of anthropogenic inadvertent tracers: organic gases and inorganic aerosols. The inorganic aerosol tracers are usually referred to as elemental tracers. Their development was stimulated first by research into the regional origins of Arctic haze and, more recently, the origins of acid rain in the eastern United States. The first regional elemental tracer was the noncrustal Mn/V ratio, which came into existence in 1980. The Mn/V ratio has now been supplanted by a seven-element system consisting of As, Se, Sb, Zn, In, and noncrustal Mn and V. In time, the seven-element system will also be expanded and improved.

Perhaps the most unappreciated aspect of elemental tracers is that they really do work and are being used regularly where other techniques fail. To date, their greatest accomplishment has been to demonstrate that (a) discrete regional signatures exist in all continental source regions studied (North America, Europe, and Asia)[1] and (b) at receptor sites, aerosol shows order-of-magnitude composition variations that are not random, but generally match the characteristics of well-known source areas. For example, we have determined that eastern North America has at least four principal regional signatures (Midwest, northeast coastal, New England/eastern Canada, and southern Ontario smelters), which can be recognized throughout the Northeast. Europe has at least four signatures (United Kingdom, western

Europe, eastern Europe, and Scandinavia), which can be followed throughout the region, particularly into the cleaner Scandinavia. In the Arctic, aerosol from the UK, Europe, and the Soviet Union can be seen clearly.

Specific conclusions drawn from elemental tracers include the following:

(a)   Aerosol from the nonferrous smelters in the Sudbury Basin of Ontario and Quebec can be detected routinely throughout New England. In spite of the large emissions of the smelters, they only seem to contribute about 2 to 3% of New England's sulfate.

(b)   Aerosol from the Midwest is seen weekly throughout the Northeast. It contributes nearly half of the Northeast's sulfate; the other half comes from the Northeast itself.

(c)   At points in the Northeast where regional apportionments of sulfate from elemental tracers can be compared with those from meteorological techniques, they seem to agree to within 10 to 20%.

(d)   The source for a particularly strong episode of black aerosol transported to Scandinavia in January 1982 was found to be eastern rather than western Europe. This conclusion was later buttressed by trajectories calculated at the Lithuanian Institute of Physics in USSR.

(e)   Aerosol from western Europe can be observed intruding into Eastern Europe whenever the winds come strongly and coherently from the west.

(f)   North America is not an important source of Arctic aerosol. Discriminant analysis of samples from both the Norwegian and Alaskan Arctic virtually never (less than 1%) identifies North American signatures there.

(g)   An unusual signature, tentatively identified as the central Soviet Union, has been found at Bear Island, Spitsbergen, and Barrow. Soviet Union sources seem to account for roughly half the Arctic aerosol during the polluted winter.

(h)   After the central Soviet Union was identified as an important source of Arctic aerosol, the transport pathway responsible for bringing it to the Arctic (via the Taymyr Peninsula) was found almost immediately.

(i)   The seasonal variation of large-scale signatures in the Arctic can be identified with, and presumably explained by, seasonal variations in large-scale meteorology leading to systematic variations in pathways.

(j)   Large-scale eddy diffusion seems to be unimportant in the short term in the Arctic, because regional signatures are discrete more often than mixed. Most aerosol is transported by organized processes (the ensemble of which amounts to eddy diffusion, however).

Development of organic gases as inadvertent anthropogenic tracers, although highly promising, has lagged behind that of elemental tracers of aerosol. Gaseous tracers are being studied primarily by R. Rasmussen and M. Khalil of Oregon Graduate Center and by M. Oehme and B. Ottar of the Norwegian Institute of Air Research. Rasmussen and Khalil have shown that methyl chloroform and Freon-22 behave much like Arctic haze and may be considered as gaseous tracers of that haze. Oehme and Ottar have demonstrated that certain chlorinated hydrocarbons show distinct differences in the Norwegian Arctic that seem to depend on whether the air has come from western or eastern Eurasia. However, both these groups need more information from the suspected source area to quantify their methods.

In spite of their successes to date, both gaseous and particulate tracers are still in early stages of development and need attention in almost every area. Foremost is the need to expand the data base at both sources and receptors. Without a full data base, the statistical reliability of the basic approach can never be properly assessed. In addition, more elements and compounds should be measured, and protocols for generating and analyzing data must be developed. Only then can these tracer systems reach their full potential.

NONCONSERVATIVE LONG-RANGE (>1000-km) ATMOSPHERIC TRACERS

Anthony Turkevich

Enrico Fermi Institute
and
University of Chicago Chemistry Department
Chicago, Illinois

Los Alamos National Laboratory
Los Alamos, New Mexico

Arguments are presented for the development and use of nonconservative (but known, less than ∿100-day lifetime) tracers for long-range (>1000-km) atmospheric experiments. The use of such tracers would prevent the buildup of background in the atmosphere, minimize the crosstalk between experiments relatively close in space or time, and make possible continuous-release experiments that would average over local micrometeorological conditions.

Examples of such tracers are $^{35}SF_6$ and various halogenated hydrocarbons containing tritium.

The purpose of this communication is to explore two concepts of long-range atmospheric tracing (two concepts that are somewhat at variance with current thinking in this area) and to propose some new tracers for use in large-scale experiments. The first concept is that the most useful tracers on this scale should <u>not</u> be conservative on more than the time-and-space scale of the phenomenon being studied. The second concept is the possible usefulness of <u>time-integrated</u> measurements of continuous tracer releases.

The elaboration of these ideas in this communication builds on the very significant accomplishments of the last few years--development and testing of two tracers (the perfluorocarbons and heavy methanes) that are being used in 1000-km-scale experiments. Certainly the highest priority should be given to exploiting the capabilities that have been demonstrated and addressing significant meteorological problems with these capabilities. However, this communication explores the probability that other tracers will be needed in the future. This is likely because of certain disadvantages in the use of the perfluorocarbons or heavy methanes as well as the desirability of studying atmospheric processes other than transport and diffusion.

The requirement that a tracer be conservative, i.e. nonreactive and relatively permanent in the atmosphere, is much too stringent for experiments that study transport on a 2- to 3-day scale (CAPTEX) or even on an intercontinental (7- to 10-day) scale. The price of using a too-long-lived tracer is the buildup of background, the ambiguity in interpretation resulting from different transport paths, and the crosstalk between experiments that take place too close in time or space. For example, an experiment that is designed to produce a signal 100 times background at a distance of 5000 km is likely to produce someplace a signal 10 times background or higher after the tagged air has been transported around the earth.

For this reason, it appears desirable to develop tracers and tracer systems that are conservative only on the time scale of interest for the phenomenon being studied. Two ways suggest themselves. First, if the tracer molecule has a radioactive atom of short half-life but is otherwise "chemically" conservative, it will not build up in the atmosphere. $^{35}SF_6$ is an example of such a tracer, although even the 85-day half-life is somewhat longer than optimal. Another approach is to take advantage of recent advances in our knowledge of atmospheric chemistry and pick molecules of relatively short chemical half-life. Examples are the various halocarbons whose half-lifes for chemical destruction in the atmosphere range from many years (e.g., CH_3CCl_3) to less than 1-year (e.g., $CHCl=CCl_2$). With one of the hydrogens substituted by tritium, these halocarbons would be very sensitive tracers.

In addition to the limited lifetime of these radioactive tracers in the atmosphere, another important advantage is the small amount that must be released. Typically, a few grams would be needed for experiments on a 5000-km scale (in contrast to the hundreds of kilograms of perfluorocarbons and kilograms of heavy methanes for similar experiments). This advantage arises from the low background levels of such radioactive tracers in the atmosphere and the sensitivity with which they can be measured. The present abundance of halocarbon molecules in the atmosphere is less than one part per billion. Although the tritium content of the compounds being considered has not been experimentally measured yet, the T/H ratio is unlikely to be even as high as 10^{-14} (the value in molecular hydrogen--a product of nuclear weapons activities) and therefore negligible. The sensitivity of tritium-counting in appropriate proportional counters is at least as low as 1 d m^{-1}, leading to a detectibility of 10^7 molecules containing tritium. Because the abundance of the halocarbons (or SF$_6$) in the atmosphere is so low, the sensitivity of a tracer experiment involving tritiated (or ^{35}S) species will be determined by the amount of air that is sampled. Starting with 100 ci, a dilution of 10^{21} will lead to 1 d m^{-1} in 4 M^3 of air. Sampling a larger amount of air will reduce correspondingly the amount of radioactivity needed.

Moreover, the "background" limitation in the case of radioactive tracers is qualitatively different from that in the case of the perfluoro- carbons or even heavy methanes. In the case of the radioactive tracers, the "background" is in the measurement equipment. The Piosson behavior of this background and of the measurement process makes possible meaningful determinations of samples having an activity level of only 10% of the background level. In addition, the measurements on a sample can usually be repeated. In contrast, the perfluorocarbon "background" is at a level in the atmosphere where variability restricts meaningful results to the order of the background level.

Working against these advantages of radioactive tracers are the disadvantages that must be either taken into account or overcome. Highest among these are the hazards of preparing and releasing multicurie amounts of radioactivity. The latter can be minimized by extended time releases (as is already necessary for other reasons for the fluorocarbons) or by focusing at the start on release points of minimal radiological impact (e.g., upper tropopause or ocean locations).

A second serious disadvantage at present is the undeveloped state of sampling and processing capabilities for such tracers. The full sensitivity potential of such tracers can be realized only by sampling large amounts of air--tens to hundreds of cubic meters. Suitable adsorbants that collect and even partially separate molecular species such as molecules of SF$_6$ either exist or can be developed. In addition, the presence of natural, or added, inactive carrier molecules can facilitate the purification process. Finally, as experiments are carried out at larger and larger scales, less extensive (time and space) sampling may be needed than for shorter range experiments. This will mean a smaller analytical load than for short-range experiments.

The second concept to be explored is the replacing the usual burst release of tracer by a continuous release that takes place over at least a season and probably several years. The sampling would now be made over a period of a month or so with sampling stations much more dispersed. The idea would be to have nature integrate over the different meteorological conditions (not requiring so much detailed meteorological data). The result would be a measurement of the <u>average</u> dilution in going from source to collector in a given season.

Although this average dilution is less closely related to micrometeorological variables, it is related to average meteorological conditions. These are usually better defined and more available without setting up special monitoring facilities. In addition, the results are more directly applicable to regional pollution problems than a small set of spot-release experiment results would be.

For such continuous-release experiments, the need for a finite life-time tracer is even more important. A half-life of 10- to 20-days in the atmosphere would be ideal for experiments on a 5000-km scale. At present, this half-life span for atmospheric tracers has not been studied extensively. However, the lifetime of a hydrocarbon in the atmosphere probably decreases with the number of hydrogen atoms present, so that halogenated compounds of the type $C_wH_yX_x$, where $w \geq 3$ and X is a halogen (e.g., F or Cl), can be chosen that have the right half-life and at the same time are easily adsorbed from large quantities of air.

The sensitivity of detection would again rest on the substitution of tritium for one of the hydrogens. The total amounts of radioactivity released would now be quite large but would be spread over many seasons. The short half-life in the atmosphere would prevent an accumulation of significant amounts of radioactivity. The sampling times would be long (tens of days) relative to those of present experiments, but once the sample were isolated from the atmosphere, the purification and measurement could proceed without time constraints.

A NEW PHYSICAL LAGRANGIAN TRACER

B.D. Zak

Sandia National Laboratories
Albuquerque, New Mexico

I. INTRODUCTION

A physical Lagrangian tracer is an airborne instrumentation system that follows the flow of air in its vicinity in both the horizontal and vertical and that can be tracked electronically. A constant volume balloon (CVB) with an appropriate payload is an approximation to a physical Lagrangian tracer. CVBs in the form of tetroons have been used extensively over the last two decades to explore many different atmospheric phenomena.[1] In the upper atmosphere, Lally[2] and others have used spherical CVBs with remarkable results. His stratospheric CVBs frequently make multiple circuits of the earth before coming down.

CVBs, however, are limited by the fact that they faithfully follow the mean flow of air in the horizontal, but not in the vertical. They follow a constant density surface, but air flows frequently do not. For many applications, the difference between an isopycnic trajectory and the actual air motion may not be significant. For long-range transport in the mixing layer on a time scale of days, the difference frequently is significant.

II. ADJUSTABLE BUOYANCY LAGRANGIAN MARKER

As part of the National Acid Preciptation Assessment Program (colloquially known as the Acid Rain Program), the Environmental Protection Agency (EPA) is developing a physical Lagrangian tracer known as the adjustable buoyancy Lagrangian marker (ABLM). The ABLM is a constant-volume balloon system whose buoyancy is adjusted to follow mean vertical flows by using air as ballast. In the ABLM, the constant-volume balloon has an inner bladder, or balloonet, which contains helium. The remaining volume of the CVB is filled with air. A system of pumps and valves allows more air to be pumped in (increasing the mean density of the balloon, causing it to go down) or to be released (decreasing the mean density, causing the balloon to go up). The idea of the ABLM was originally proposed by Lally[2] in 1967. The heart of the ABLM is a set of sensors serviced by a microprocessor. The inputs from the sensors allow the microprocessors to determine if a buoyancy adjustment is needed.

The micrprocessor can be programmed to use a number of different algorithms to control the ABLM buoyancy. Since air motion aloft is nearly isentropic, an isentropic algorithm is an appropriate choice. Alternately, if relative vertical air motion can be measured sufficiently accurately, an algorithm can be constructed to keep the cumulative relative vertical air motion near zero. However, the demancs on the accuracy of the relative vertical air motion sensor must be quite stringent for a marker to be useful over a period of days.

III.   REQUIREMENTS AND SPECIFICATIONS

The requirements that the EPA has placed on the ABLM:

- Lifetime $\geq$ 3 days
- Tracking range > 1000 km
- System capable of handling several (10 to 50) markers at a time
- Cheap enough to be disposab_e
- Follows mean vertical flows with "acceptable" fidelity

A "testbed" prototype system built to demonstrate feasibility meets many but not yet all of the requirements (Fig. 1). After feasibility has been established, a "production" prototype wi_l be developed which meets all requirements. On the basis of the testbed prototype, the likely approximate specifications of the ABLM can be discerned. They include:

- Balloon volume: 12 to 15 m^3
- Balloon diameter: ~3 m
- Payload: 3 to 4 kg distributed in 2 to 3 packages
- Sensors:
    - Relative vertical air motion
    - Pressure
    - Temperature
    - Wet bulb temperature
    - Superpressure
    - Balloon skin strain
- Tracking and Data Handling ɔy an ARGOS system on TIROS satellites

IV.   CURRENT STATUS

Tests on the testbed prototype ABLM are just beginning in a benign environment. The site of the tests is a 70-m-high tower associated with a

Fig. 1. The 3-m-diam. balloon to be used as part of the ABLMs

solar central receiver test facility at Sandia National Laboratories. The tower offers a 10-m-square by 52-m-high enclosed volume in which the tests are taking place. Preliminary data suggest that the concept is sound. Later this summer tests will be conducted in the open atmosphere.

V.   PREVIOUS EXPERIENCE WITH QUASI-LAGRANGIAN SYSTEMS

In the past, quasi-Lagrangian systems have been used to track plumes for long distances. In project DaVinci,[3,1] for instance, the position of a manned zero-pressure balloon relative to the centerline of a plume was measured over more than 12 hours and 300 km downwind. To within experimental error (±6 km), the relative position of the plume centerline and the balloon remained fixed, but here the transport occurred mostly during the day when a minimum of wind shear could be expected.

In the Tennessee Plume Study,[5] a superpressure Lagrangian measurement platform was inserted into the plume from the Cumberland power plant. The experiment continued for 8 hours during which the balloon was carried 160 km downwind. The plume was tracked by aircraft measurements of the balloon and by onboard-balloon measurements of the concentration of the SF_6 tracer that was being released in the stack of the Cumberland plant as the balloon lifted off.

VI.   SUMMARY

Although much work remains to be done, it now appears that a physical Lagrangian tracer will be operational and available for use within the near future. The tracer is an adjustable-buoyancy constant-volume balloon with onboard machine intelligence and an appropriate array of sensors. Worldwide tracking and data reporting will be accomplished by means of the ARGOS satellite-borne data system.

REFERENCES

1.   Zak, B. D. (1983) "Lagrangian Studies of Atmospheric Pollutant Transformations," in Trace Atmospheric Constituents: Properties, Transformation, and Fates, edited by S. E. Schwartz. Wiley, New York.

2.   Lally, V. E. (1967) Superpressure Balloons for Horizontal Soundings of the Atmosphere, National Center for Atmospheric Research Technical Note NCAR-TN-28.

3.   Zak, B. D. (1981) Final Report on Project DaVinci: A Study of Long Range Air Pollution Using a Balloon-Borne Lagrangian Measurement Platform, Vol. 1: Overview and Data Analysis. Sandia National Laboratories Report SAND78-0403/1.

4.   Zak, B. D. (1981) "Lagrangian Measurements of Sulfur Dioxide to Sulfate Conversation Rates," Atmosph. Environ. 15, 2583.

5.   Gay, G. T. et al. (1981) Lagrangian Measurement Platform Flight in Support of the Tennessee Plume Study: Field Effort and Data, Sandia National Laboratories Report SAND79-1336.

Appendixes

APPENDIX A—ORGANIZING COMMITTEE

Dr. David S. Ballantine
Office of Health and Environmental Research
U.S. Department of Energy

Dr. Sumner Barr
Los Alamos National Laboratory

Dr. William E. Clements
Los Alamos National Laboratory

Dr. Russell N. Dietz
Brookhaven National Laboratory

Dr. Paul Gudiksen
Lawrence Livermore Laboratory

Dr. Paul R. Guthals
Los Alamos National Laboratory

Dr. Jerome L. Heffter
National Oceanic and Atmospheric Administration

Dr. Thomas Horst
Pacific Northwest Laboratories

Dr. Richard Semonin
Illinois State Water Survey

APPENDIX B—MEMBERS OF DISCUSSION GROUPS

<u>APPLICATIONS OF TRACERS TO</u>
<u>ACID PRECIPITATION</u>

Jake Hales, PNL - Chairman

Dr. Arthur Bass
Environmental Research and Technology, Inc.

Dr. Ernest A. Bondietti
Oak Ridge National Laboratory

Dr. Julius Chang
National Center for Atmospheric Research

Dr. Warren Johnson
SRI International

Dr. Raymond J. Lagomarsino
USDOE/Environmental Measurement Laboratory

Dr. Howard Liljestrand
University of Texas

Dr. B. B. McInteer
Los Alamos National Laboratory

Dr. Paul Michael
Brookhaven National Laboratory

Dr. Kenneth A. Rahn
University of Rhode Island

Dr. Steve Schwartz
Brookhaven National Laboratory

Dr. Richard Semonin
Illinois State Water Survey

Dr. Peter Summers
Atmospheric Environment Service

Dr. M. Wesely
Argonne National Laboratory

Dr. W. Wilkes
Mound Laboratories

CONTINENTAL- AND GLOBAL-SCALE TRACERS AND THEIR APPLICATION

Dr. Bernard Zak, SNL - Chairman

Mr. Robert Baxter
Airovironment, Inc.

Dr. Rudolf Engelmann
National Oceanic and Atmospheric Administration

Dr. Jerome L. Heffter
NOAA- Air Resources Laboratories

Dr. Stuart Kupferman
Sandia National Laboratories

Dr. Robert Malone
Los Alamos National Laboratory

Dr. John Merrill
University of Rhode Island

Dr. Eugene J. Mroz
Los Alamos National Laboratory

Dr. Joseph M. Prospero
University of Miami

Dr. Kenneth A. Rahn
University of Rhode Island

Dr. Fred Shair
California Institute of Technology

Dr. Anthony Turkevich
University of Chicago

Dr. Allen Weber
Savannah River Laboratory

Dr. Tetsuji Yamada
Los Alamos National Laboratory

DRY DEPOSITION AND RESUSPENSION

Tom Horst, PNL - Chairman

Dr. J. Christopher Doran*
Pacific Northwest Laboratories

Dr. Ekkehard Dreiseitl
Institute for Meteorology and Geophysics, University of Innsbruck

Dr. John Evans
Pacific Northwest Laboratories

Dr. John Garland*
Environmental and Medical Sciences Division, AERE Harwell

Dr. L. Husain
New York State Department of Health

Dr. William R. Kelly*
National Bureau of Standards

Dr. Leonard Newman
Brookhaven National Laboratory

Mr. William Ohmstede
US Army Atmospheric Sciences Laboratory

Dr. W. G. N. Slinn*
Pacific Northwest Laboratory

Dr. Lynn Teuscher*
Tracer Technology of California

* Responsible for written discussion summary

REGIONAL AIR QUALITY

Frank Gifford, Los Alamos Consultant - Chairman

Dr. William E. Clements
Los Alamos National Laboratory

Dr. Marvin H. Dickerson
Lawrence Livermore National Laboratory

Dr. Russell N. Dietz
Brookhaven National Laboratory

Mr. Gilbert J. Ferber
National Oceanic and Atmospheric Administration - Air Resource Laboratories

Dr. P. Gaglione
Radiochemistry Division, Joint Research Centre, Italy

Dr. D. Gillette
National Oceanic and Atmospheric Administration

Dr. Glenn Gordon
University of Maryland

Dr. Philip W. Krey
USDOE/Environmental Measurement Laboratory

Dr. Yeng-Hwa Kuo
National Center for Atmospheric Research

Dr. R. Leaitch
Atmospheric Environment Service

Dr. John Ondov
Martin-Marietta Environmental Division

Dr. Montie Orgill
Pacific Northwest Laboratories

Mr. Donald Shearer
TRC - Denver

Dr. C. David Whiteman
Pacific Northwest Laboratories

NATURAL SOURCES OF INORGANIC AND ORGANIC
GASES AND AEROSOLS

Jeff Gaffney, BNL - Chairman

Dr. Sumner Barr
Los Alamos National Laboratory

Dr. Donald F. Gatz
Illinois State Water Survey

Dr. Paul Gudiksen
Lawrence Livermore National Laboratory

Dr. Bruce B. Hicks
Atmospheric Turbulence and Diffusion Laboratory

Dr. George A. Sehmel
Pacific Northwest Laboratories

Dr. Hanwant B. Singh
SRI International

Dr. Eugene Start
National Oceanic and Atmospheric Administration/ARL-ID

Dr. Chris Walcek
National Center for Atmospheric Research

Dr. Marvin Wilkening
New Mexico Tech.

Dr. William Zoller
Los Alamos National Laboratory

APPENDIX C—WORKSHOP PARTICIPANTS

Dr. David S. Ballantine
Office of Health and Environmental Research
Physical and Technological Research Division
ER-74, GTN
US Department of Energy
Washington, DC 20545

Dr. Sumner Barr
Los Alamos National Laboratory
P.O. Box 1663, MS D466
Los Alamos, NM 87545

Mr. Robert Baxter
Airovironment, Inc.
145 Vista Avenue
Pasadena, CA 91107

Dr. Ernest A. Bondietti
Oak Ridge National Laboratory
Environmental Sciences Division
P.O. Box X, Building 1505
Oak Ridge, TN 37831

Dr. Julius S. Chang
National Center for Atmospheric Research
P.O. Box 3000
Boulder, CO 80307

Dr. William E. Clements
Los Alamos National Laboratory
P.O. Box 1663, MS D466
Los Alamos, NM 87545

Dr. Walter Dabbert
SRI International
333 Ravenswood Avenue
Menlo Park, CA 94025

Dr. Marvin H. Dickerson
Lawrence Livermore National Laboratory
P.O. Box 808
Livermore, CA 94550

Dr. Russell N. Dietz
Environmental Chemistry Division
Brookhaven National Laboratory
Building 426
Upton, NY 11973

Dr. J. Christopher Doran
Pacific Northwest Laboratories
P.O. Box 999
Richland, WA 99352

Dr. Ekkehard Dreisteitl
Institute for Meteorology and Geophysics
University of Innsbruck
Schopfstrasse 41
A-6020 Innsbruck
AUSTRIA

Dr. Rudolf Engelmann
National Oceanic and Atmospheric Administration
6010 Executive Boulevard
Rockville, MD 20852

Dr. John Evans
Pacific Northwest Laboratories
P.O. Box 999
Richland, WA 99352

Mr. Gilbert J. Ferber
NOAA – Air Resources Laboratories
6010 Executive Boulevard (WSC-5)
Rockville, MD 20852

Dr. Jeff Gaffney
Brookhaven National Laboratory
Upton, NY 11973

Dr. P. Gaglione
Radiochemistry Division
Joint Research Centre
21020 Ispra Varese
ITALY

Dr. John Garland
Environmental and Medical Sciences Division
AERE Harwell, Building 364
Oxfordshire, OX11 ORA
ENGLAND

Dr. Donald F. Gatz
Illinois State Water Survey
P.O. Box 5050, Station A
Champaign, IL 61820

Dr. Franklin A. Gifford
109 Gorgas Lane
Oak Ridge, TN 37830

Dr. D. Gillette
National Oceanic and Atmospheric Association
325 Broadway
Boulder, CO 80301

Dr. Henry Goldwire
Lawrence Livermore National Laboratory
P.O. Box 808, L-451
Livermore, CA 94550

Dr. Glenn Gordon
Department of Chemistry
University of Maryland
College Park, MD 20742

Dr. Paul Gudiksen
Lawrence Livermore National Laboratory
P.O. Box 808
Livermore, CA 94550

Mr. Paul R. Guthals
Los Alamos National Laboratory
P.O. Box 1663, MS J514
Los Alamos, NM 87545

Dr. Jeremy M. Hales
Atmospheric Chemistry Section
Atmospheric Sciences Department
Pacific Northwest Laboratory
P.O. Box 999
Richland, WA 99352

Dr. Jerome L. Heffter
NOAA - Air Resources Laboratories
6010 Executive Boulevard (WSC-5)
Rockville, MD 20852

Dr. Bruce B. Hicks
Atmospheric Turbulence and Diffusion Laboratory
P.O. Box E
Oak Ridge, TN 37830

Dr. Tom Horst
Pacific Northwest Laboratories
P.O. Box 999
Richland, WA 99352

Dr. L. Husain
DIR, Chemical Sciences Laboratory
Environmental Health Institute
New York State Department of Health
Empire State Plaza
Albany, NY 12201

Dr. Warren Johnson
Aerophysics Laboratory L2001
SRI International
333 Ravenswood Avenue
Menlo Park, CA 94025

Dr. William R. Kelly
Center for Analytical Chemistry
National Bureau of Standards
Washington, DC 20234

Dr. Philip W. Krey
Director, Analytical Chemistry Division
USDOE/Environmental Measurement Laboratory
376 Hudson Street
New York, NY 10014

Dr. Yeng-Hwa Kuo
National Center for Atmospheric Research
P.O. Box 3000
Boulder, CO 80307

Dr. Stuart Kupferman
Seabed Programs Division
Organization 6334
Sandia National Laboratories
Albuquerque, NM 87185

Dr. Raymond J. Lagomarsino
USDOE/Environmental Measurement Laboratory
376 Hudson Street
New York, NY 10014

Dr. R. Leaitch
Atmospheric Environment Service
4905 Dufferin Street
Downsview, Ontario M3H 5T4
CANADA

Dr. Howard Liljestrand
Department of Civil and Environmental Engineering
University of Texas
Austin, TX 78712

Dr. Robert Malone
Los Alamos National Laboratory
P.O. Box 1663, MS F665
Los Alamos, NM 87545

Dr. B. B. McInteer
Los Alamos National Laboratory
P.O. Box 1663, MS J568
Los Alamos, NM 87544

Dr. John Merrill
Center for Atmospheric Chemistry
Graduate School of Oceanography
University of Rhode Island
Kingston, RI 02881

Dr. Paul Michael
Atmospheric Sciences Division
Department of Energy and Environment
Brookhaven National Laboratory
Upton, NY 11973

Dr. Eugene J. Mroz
Los Alamos National Laboratory
P.O. Box 1663, MS J514
Los Alamos, NM 87545

Dr. Leonard Newman
Brookhaven National Laboratory
Upton, NY 11973

Mr. William Ohmstede
US Army Atmospheric Sciences Laboratory
ATTENTION: DELAS-BE-C WSMR, New Mexico 88002

John Ondov
Martin-Marietta Environmental Division
9200 Rumsey Road
Columbia, MD 21045

Dr. Montie Orgill
Pacific Northwest Laboratories
2400 Stevens
Richland, WA 99352

Dr. Joseph M. Prospero
School of Marine and Atmospheric Science
University of Miami
4600 Rickenbacker Causeway
Miami, FL 33149

Dr. Kenneth A. Rahn
Associate Research Professor
Graduate School of Oceanography
University of Rhode Island
Kingston, RI 02881

Dr. Elmar Reiter
Department of Atmospheric Science
Colorado State University
Fort Collins, CO 80521

Dr. Steve Schwartz
Brookhaven National Laboratory
Associated Universities
Upton, NY 11973

Dr. George A. Sehmel
Pacific Northwest Laboratory
2400 Stevens, 1100 Area
Richland, WA 99352

Dr. Richard Semonin
Illinois State Water Survey
Box 5050, Station A
Champaign, IL 61820

Dr. Fred Shair
California Institute of Technology
1201 E. California Road
Pasadena, CA 91125

Mr. Donald Shearer
TRC-Denver
8775 E. Orchard Road
Englewood, CO 80111

Dr. Hanwant B. Singh
SRI International
333 Ravenswood Avenue
Menlo Park, CA 94025

Dr. W. G. N. Slinn
Atmospheric Sciences Department
Pacific Northwest Laboratory
P.O. Box 999
Richland, WA 99352

Dr. Eugene Start
NOAA/ARL-ID
550 2nd Street
Idaho Falls, ID 83401

Dr. Peter Summers
Atmospheric Environment Service
4905 Dufferin Street
Downsview, Ontario M3H5T4
CANADA

Dr. Lynn Teuscher
Tracer Technology of California
2120 W. Mission Road
Ecandido, CA 92025

Dr. Anthony Turkevich
Enrico Fermi Institute
University of Chicago
Chicago, IL 60637

Dr. Chris Walcek
National Center for Atmospheric Research
P.O. Box 3000
Boulder, CO 80307

Dr. Allen H. Weber
Savannah River Laboratory
Building 773-A, A1019
Aiken, SC 29808

Dr. M. Wesely
Atmospheric Physics Section
Argonne National Laboratory
Argonne, IL 60439

Dr. C. David Whiteman
Pacific Northwest Laboratories
P.O. Box 999
Richland, WA 99352

Dr. Marvin Wilkening
Professor of Physics
New Mexico Tech.
Socorro, NM 87801

Dr. W. Wilkes
Mound Laboratories
Box 32
Miamisburg, OH 45342

Dr. Tetsuji Yamada
Los Alamos National Laboratory
P.O. Box 1663, MS D466
Los Alamos, NM 87545

Dr. Bernard Zak
Applied Atmospheric Research Division
Organization 6324
Sandia National Laboratories
Albuquerque, NM 87185

Dr. William Zoller
Los Alamos National Laboratory
P.O. Box 1663, MS J514
Los Alamos, NM 87545

Other Noyes Publications

FUTURE ATMOSPHERIC CARBON DIOXIDE SCENARIOS AND LIMITATION STRATEGIES

by

J.A. Edmonds, J. Reilly
Institute for Energy Analysis

S. Seidel
U.S. Environmental Protection Agency

D. Rind, S. Lebedeff
NASA Goddard Space Flight Center

D. Keyes
Tucson, Arizona

J.P. Palutikof, T.M.L. Wigley, J.M. Lough
University of East Anglia, U.K.

M. Steinberg
Brookhaven National Laboratory

J.R. Trabalka, D.E. Reichle, T.J. Blasing, A.M. Solomon
Oak Ridge National Laboratory

This book discusses various climatic scenarios relating to the future buildup of carbon dioxide levels in the atmosphere, as well as various options and limitations strategies for controlling or slowing the concentrations of carbon dioxide levels in the atmosphere.

Although projections for future CO_2 concentrations in the atmosphere are fairly straightforward, depending upon the figure used for energy growth rates based on fossil fuels, there are other factors that make predictability very uncertain. For example: the extent that nonfossil fuels will be utilized; the extent of future deforestation and land clearing; a determination of how emissions will be partitioned among the atmosphere, the oceans, and the biosphere; the effects on climate of other gases released from industrial activities; etc.

The book is presented in six parts, which consider the CO_2 problem from varying aspects. A condensed table of contents listing **part and chapter titles** is given below.

ISBN 0-8155-1064-0 (1986)

620 pages

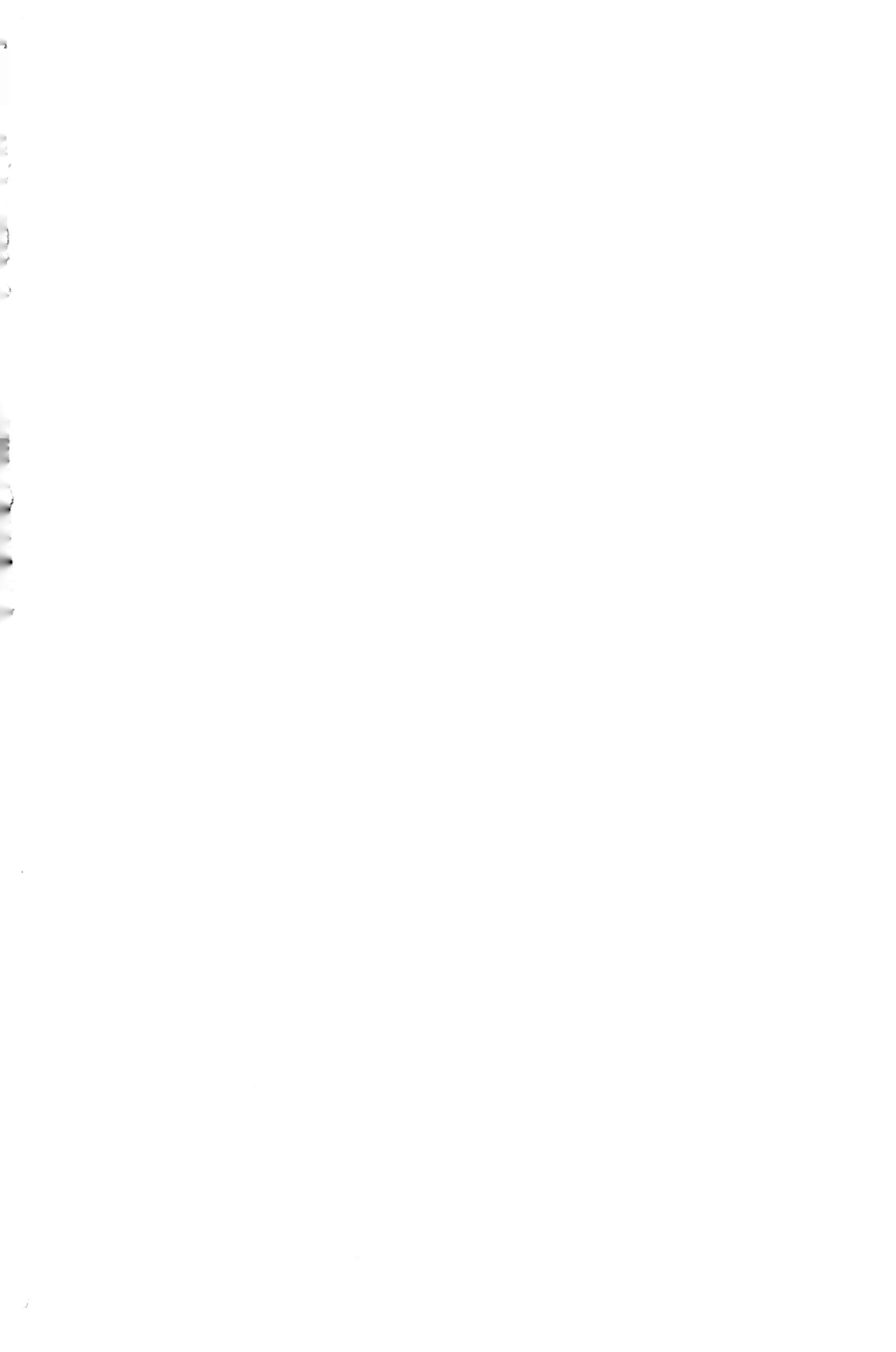